BARCODE IN BACK

HUMBER LIBRARIES LAKESHORE CAMPUS
3199 Lakeshore Blvd West
TORONTO, ON.   M8V 1K8

Prologue

It is with great pride that ICO presents to you noAH-8, the newest and most exciting edition in our exclusive series of package design annuals.

International Creators' Organization has been a respected leader in the world of creative design for over 30 years. We are passionate in our goal to provide a high quality meeting place for talented creators and distinctive clients worldwide.

noAH-1, the first book in our package design series, was published in March of 1985. It was created as a way to showcase the artwork of many outstanding designers from around the world. Our hope was to provide these talented artists with a stage upon which they could present their work, and to give them an opportunity to be discovered by international clientele. Thus noAH-1 was born, and began its maiden voyage around the world.

Since ICO's first package annual of noAH-1 was launched, others have followed. The sails of our fleet of noAH ships are fully expanded now, and we have traveled to all of the world's market places. ICO sincerely appreciates the warm reception and enthusiastic support we have received for our series of noAH books over the years. Our mission is to continue our search for excellence in the field of package design, and to share it with you in future noAH publications.

Creative treasures from over 27 countries are piled up and waiting for you in noAH-8. Please join us now in discovering the outstanding artwork of the talented designers presented in these pages. We hope you will enjoy the adventure.

International Creators' Organization ICO
Jo Sickbert

HUMBER LIBRARIES LAKESHORE CAMPUS
3199 Lakeshore Blvd West
TORONTO, ON.   M8V 1K8

Graphic Design Technology

Almost two years ago I decided instead of moving into a new office space I was going to set up a home office and run my business from there. I was a little bit skeptical about it. If it didn't work out at least I didn't waste any money on renting a new office. I don't regret my decision a single bit. I really enjoy working at home. My commute is less than 5 minutes from bed to desk. I'm helping to conserve the environment by driving less, except for the occasional golf outing.

Computer technology has become much more advanced, especially for communication. Now my assistants can stay and work from home and we can connect over the network to communicate with each other. A few years ago this was not the case, before the technology became what it is today, designers staying at home doing freelance work could not have a full career under the existing conditions. Working from the home was only a temporary part time job. Nowadays we have clients in Atlanta, Boston, Chicago, overseas and locally in San Francisco. We physically never go to see the client to do any presentation or coordination. No matter what type of project, corporate identity, brochures or annual reports we send PDFs for presentation and approval. What once took several days goingback and forth can be done in less than one workday.

All my generation colleagues are retired except for Barry Deustch and Tom Ingalls both who are featured in this book. My generation experienced a lot of frustration with the current attitudes towards the industry. Some clients do not know or see the difference between graphic design and desktop publishing. We as designers had to drastically change the way we work in order to stay in business.

Even though desktop publishing has cut into the industry, packaging design is still strong. Being able to use design software does not a package designer make.

This year the economy is coming back. We're past the turning point in the industry and moving toward a more positive future.

A couple of years ago I seriously considered retiring. Setting up the home office has made my life easier and more fun. I can still work in graphic design, something I'd love to do for the rest of my life, even after 30 years in the business. The home office has allowed me to create my own time to play golf and enjoy time with the girls, my three beautiful cats and the wife. Thank you for the technology!

Yashi Okita
Yashi Okita Design

Brand design in Russia point of view

Russian market of brand-design services is rather young due to political and economical reasons. After the dissolution of USSR in 1991 the situation on Russian consumer market changed dramatically. The country tried to catch up the rest of the world.

I suppose that brand design in Russia originated in late 90-ies. After so called economical "crisis" in 1998 the devaluation of local currency cut imports for many foreign producers of consumer products. Local companies started to produce more in order to satisfy the rising demand on FFCG market. They should have delivered the products in modern packaging that could compete with foreign brands. Packaging design and new brands creation were in great demand that time. Depot WPF was the first local agency founded in 1998 that offered to its clients the services on packaging design and brand development.

Today the Russian market is an important and attractive one for major Western companies and brand design in Russia is very perspective and fast developing industry. Talented Russian designers create brands not only for local but also for European markets. They receive the highest awards on international contests and festivals competing with their colleagues from all over the world.

There are a lot of challenges for the designers. In the last few years the life and as a consequence the economic reality has been changed. All processes have been moving and progressing very rapidly. Today's consumer is freakish and indulged. The produces try to satisfy all consumer whims in different displays and present leading edge products. They also stand for short-term brand campaigns. There is no use to carry out the burden of obsolete values - just renew a brand and go further!

Sometimes it is more effective for the brand to bring it back to the launch stage than to breathe life into iconic name.

Anna Lukanina
Depot WPF Brand and Identity

Towards More Perfect Packaging Design

Designers' design careers are like the process of life, or a process of catharsis. Designers must evolve from childishness to maturity, from ignorance to knowing, from thoughtlessness to wisdom, and from sin to compassion. The designer is constantly growing and engaging in self-improvement; and is always striving to make progress towards better and more perfected design.

Young designers find it next to impossible to establish their own personal styles, heed their preferences, or pursue their ideals. Young designer are also forced to indulge in superficiality and exaggeration, and must design many impractical, costly, and over-packaged works. However, as they accumulate experience and knowledge and begin to think deeply, designers can eventually learn to achieve the best effects in the shortest time with the least consumption of resources.

There's no need to use two lines when a single line is more effective. There's no need to use ten patches of color when a single patch of color will have more emotional impact. When economically-efficient, practical materials can achieve the desired goal, there's no need to make lavish use of expensive materials. Achieving effective design depends on the designer's experience, wisdom, and vision. Sumptuous packaging is typically used only once and then discarded; we often see cast-off piles of expensive packaging materials. This kind of packaging conveys a false, deceitful brand image, and it is devoid of sincerity, thoughtfulness, or wisdom.

Designers must undergo continuous learning and accumulation of experience before they can uncover the true principles that are hidden in different shapes. Shapes have their own distinctive characteristics, inner feelings, and stylistic timeliness. A designer must spend a lifetime becoming friends with shapes. The designer must team up with shapes for the sake of adventure, hard work, enjoyment of life, and overcoming difficulties. Only then can the designer thoroughly understand shapes, and can use a point, a line, or a patch of color to the fullest extent. Only then can the designer impress the world and move the hearts of millions of people. Only then can the designer create new visual states of harmony and perfection, and help bring an endlessly wonderful world into existence. This kind of packaging design is truly an art.

Prof. Jeffrey Su
National Taiwan Normal University

the WOW factor

Before I commenced writing this essay, I looked up the word 'package' in my reasonably comprehensible dictionary and found the following description, 'A wrapped or boxed object'.

Packaging design, of course, is not rocket science but creative skills are required of a packaging designer to create a successful package. Packaging design is much more than my dictionary's simplistic description.

To be a successful package design here are my criteria:

1. Firstly yes, the package must be stylish in form and have functional relevance for the product.
2. The design must be based on fact and display the product brand name/identity, relevant descriptive terminology and any unique or quality points of interest.
3. All mandatory copy must be presented clearly and precisely in required legal type sizes.
4. The package design must project and present the product's unique personality, characteristics and special qualities in an attractive, exciting and/or stunning fashion.
5. The packaging design must, above all things, attract consumers to the product and influence the consumers to purchase and take the product home with them.

The question is then, 'How does a designer create successful package design that not only creates attention but sells?' The answer is to carry out all the points I outlined in my list of criteria - but with the important addition of an IDEA!

The designer must know all about the product and the creator/s of the product to formulate an idea. Creating great ideas comes from practice.

Think of the brain as a muscle. The more you use it, train it to think, the better the thinking and creative performance will be! Without a great idea, a design cannot possibly succeed. The idea gives the concept life.

Having put a great idea into motion, a designer can then add the 'MAGIC' - it might be the method of presentation, for example, creative use of colour, an illustrative or photographic style, a printing method or material choice which gives the finished package presentation the clarity, excitement or sophistication to achieve the important point-of-difference from others in the competitive market place to succeed.

It is called the 'WOW' factor!
It is packaged product success - by design.

Barrie Tucker

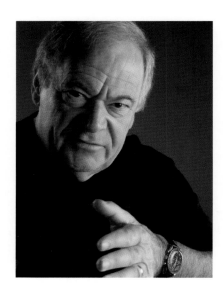

How does a pack tell a great story?

Storytelling is a timeless art and a powerful way of connecting with people. More than ever brand owners see their packaging as a great platform to tell their story. But not all stories are compelling. Some are tedious, unbelievable or just irrelevant.

Indeed in packaging design there is a danger of story-telling becoming one-dimensional, limited to a series of charming but rather predictable anecdotes. Sometimes the story tries to say too much, losing interest and clarity. Or worse still the story becomes someone else's story, lifted from another brand.

Yet packaging design has the power to tell a story that is so much more evocative.

Innocent is a UK brand that tells a very original story in a fresh, distinctive and resonant tone of voice. So much so that it has spawned a generation of Innocent wanabees who have tried to adopt the Innocent tone of voice as their own. And there are no geographical boundaries -Do Bem in Brazil and Nudie in Australia both have more than a passing resemblance to Innocent.

But the moment you adopt someone else's story you weaken your own. You lose authenticity and credibility. At the heart of a compelling story is a universal truth and all brand stories should focus on their fundamental truth. Otherwise the story simply becomes at best an entertaining distraction.

There are many ways to tell a story. And the best way will be your way.
Humorous and anecdotal may be right for one brand, factual and informative may be right for another and symbolic and suggestive for the next. Yet whole categories can become seduced and blinkered by a certain genre - whiskies love the anecdote of history, perfumes love the rhythm of poetry and skincare loves the authority of science.

Memorable brand stories are the ones that are able to break out of the category style. But more than that they are able to develop an overall narrative across everything they do. It results in a story that's not just about words or tone of voice but is able to tell a much more holistic and coherent story.

Nude, a new organic biodynamic skincare range, is a good example of this. Nude's packaging tells a very sensual and stripped down story where a perfectly balanced logotype, pebble-like bottles shapes, minerealesque colours, sustainable materials and only essential copy all evoke a highly evocative story. Yet the packaging doesn't attempt to tell every chapter of that story. Some things are left to the imagination and some things are left to other media.

And other brands approach their stories in very different ways. Coca Cola Blak takes an iconic story into the territory of coffee through a powerful and rich interpretation of the brand's key graphic equities, Union Hand Roasted expresses the story of hand-crafted ethical coffee through bold, raw symbolism, Dr Stuart's brings a wonderful English eccentricity to herbal tea through humorous illustrations and a very new brand of rooibos herbal tea Kromland Farm cues the spirit and soul of South Africa through its fresh naive typeface and illustrative style.

Words are an increasing part of packaging design - yet at the same time stories are not just about words. It's a mistake to see the brand story as the back label copy - or even the front label copy. The whole pack tells a whole story - and as such draws on the full palette of graphics, typography, words, materials, shape and texture to bring that story to life in a desirable way. This may seem rather obvious - but if it is so obvious then why don't more brands do it?

Mike Branson
Managing Partner
Pearlfisher

Aesthetic Perception of Package Design

Perception involves the mind acting upon matter to organize stimuli into recognizable forms. On the simplest level, to perceive is to identify. But perception is more than just a collection of sensory information. It involves problem solving, analysis and synthesis.

Package design employs unique symbol systems for the communication of ideas. These symbols are not substitutes; they portray something more abstract. Symbols are human inventions for the mediation of the self and the world. Such inventions fall into distinct categories.

Conventional symbols are arbitrary and culture-bound. Representational symbols are imitative, designed to represent a specific aspect of reality. Connotative symbols are distorted to emphasize particular qualities of the referent. Qualitative symbols are designed to represent an idea or feeling. Outstanding examples of each of these kinds of symbols are evident in the images featured in this edition. The reader is invited to take the time to use this framework to appraise and comprehend each unique representation. The delight in metacognitively appreciating what each image communicates will be worth the effort.

Faith Dennis Morris Ed.D

JOÃO MACHADO
PORTUGAL

João Machado was born in Coimbra, 1942. Graduated in Sculpture by the Oporto Fine School of Arts.

Individual Exhibitions

1986	Art Poster Gallery, Lambsheim, Germany
1987	Annecy/Bonlieu - Centre d'Action Culturelle, France
1989	Lincoln Center, Colorado, EUA
1996	Galeria de la Casa del Poeta, Mexico
1997	DDD Galery, Osaka, Japan
1998	Casa Garden, Macau
2001	Pécsi Galéria, Pécs, Hungary
2002	Dansk Plakatmuseum, Arhus., Germany
2006	Ginza Graphic Gallery, 20th Anniversary GGG / DDD Project, Japan.
2007	International Triennial of Stage Poster, Bulgary.

Awards

1983	1st Award "Prémio Nacional Gulbenkian para a melhor Ilustração de Livros para a Infância", Portugal
1989	Special Award "Die Erste Internationale Litfass Kunst Biennale", Germany Bronze Medal "Bienal do Livro de Leipzig", Germany
1996	Award, "Computer Art Bienal", Rzeszow, Poland
1997	1st Award "Mikulás Galanda, Bienal do Livro de Martin", Slovak Republic; 1st Award "First International Competition for Fair Poster", Bulgary
1997	1st Award "Logo Film Commission, Association of film Commissioners International Denver", USA
1999	1st Award "Best of Show, European Design Annual", Great Britain; Award Zgraf 8 Icograda Excellence, Croatia
2004	2nd Award, "4th International Triennal of Stage Poster", Bulgary.
2005	Award "Aziago International Award 2005", for the best worldwide stamp in the Tourism category, Italy

1
1. João Machado
2. Chávena / Cup / Tasse * Café / Coffee / Kaffeee
3. Package for a cofee cup
4. A lath made of brass covered with paper Tintoreto 250grs.
5. This all piece (package and cup illustration) has been created for merchandising.

2
1. Câmara Municipal de Guimarães
   (Guimarães City Hall) – Guimarães, Portugal
2. Guimarães – Património Cultural da Humanidade
   (Guimarães, World Heritage)
3. Package for a Bronze Medal
4. Paper Tintoretto 360grs.
5. This piece has been created for Guimarães City Hall once
   the historic centre of Guimarães has been classified has
   world heritage in 2001.

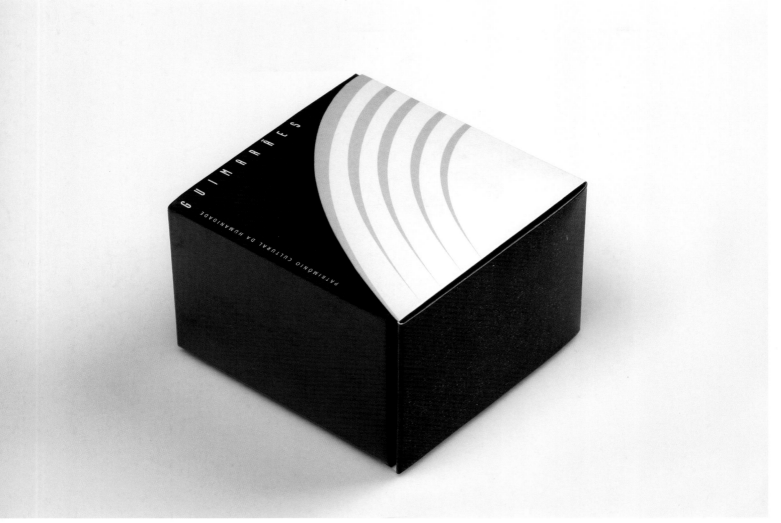

3   1. Câmara Municipal de Almada
   (Almada City Hall) – Almada, Portugal
2. Coffee cup
3. Package for a Coffee cup
4. Papers: Trucard 400grs. and Notturno (Black) 450gs.
5. This all piece (package and cup illustration) has been created for the Presidency of the City Hall to use it as a Christmas Gift.

4 1. Câmara Municipal de Almada
     (Almada City Hall) – Almada, Portugal
  2. Tea cup
  3. Package for a tea cup
  4. Papers Trucard 400grs. and Notturno (Black) 450gs.
  5. This all piece (package and cup illustration) has been
     created for the Presidency of the City Hall
     to use it as a Christmas Gift.

INGALLS DESIGN
U.S.A.

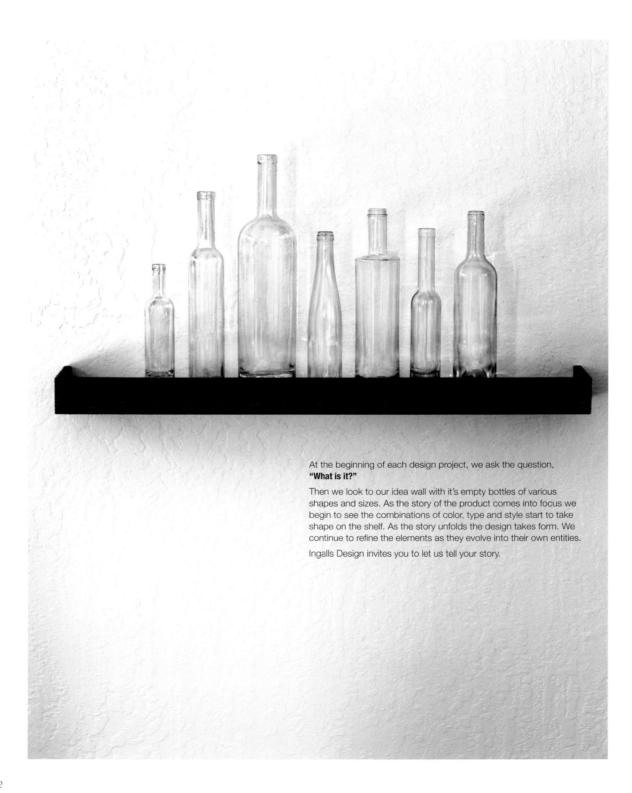

At the beginning of each design project, we ask the question, **"What is it?"**

Then we look to our idea wall with it's empty bottles of various shapes and sizes. As the story of the product comes into focus we begin to see the combinations of color, type and style start to take shape on the shelf. As the story unfolds the design takes form. We continue to refine the elements as they evolve into their own entities.

Ingalls Design invites you to let us tell your story.

1.
Ingalls Design
Promotional Wine
Anderson Valley Merlot
Glass, printed paper
Sent with brochure

2.
JAX Vineyards
Y3 Wines
Chardonnay
Glass, printed paper
High end white wine

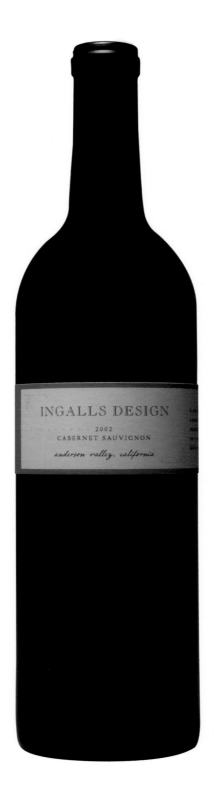

3.
St. George Spirits
Agua Perfecta
Poire Liqueur
Glass, printed paper
Pear william eau de vie

4.
Craft Distillers
Germain-Robin
Aged select brandy
Glass, printed paper
California Brandy

5.
Daniel Gehrs Wines
Daniel Gehrs
Eclipse Pinot Noir
Glass, printed paper
Vintage California Pinot

6.
Dale Family Vineyards
Ringside
Cabernet Sauvignon
Glass, printed paper
California Cabernet

5.

6.

7.
St. George Spirits
Hanger One
American Vodka
Silkscreen on glass
Premium Vodka

8.
Daniel Gehrs Wines
Daniel Gehrs
Vintage Port
Glass, printed paper
Vintage California Port

9.
Napa Nevada Spirits
Prime Rib Red
California Red
Glass, printed paper
Hearty California red wine

10.
Napa Nevada Spirits
Vince Vineyards
Cabernet Sauvignon
Glass, printed paper
Vintage Cabernet

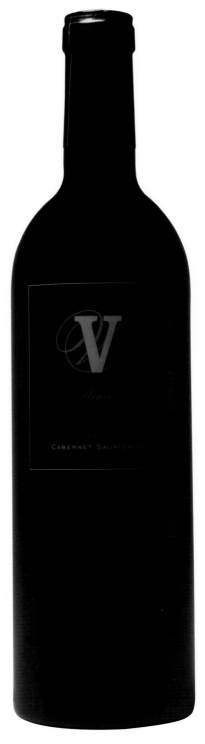

11.
Fox River Paper
Estate Label Paper
California Merlot
Glass, foil, paper, varnish
Label paper promotion

12.
Bronco Wines
Pacific Oasis
Merlot
Glass, printed paper
Vintage California Merlot

13.
Greenfield Wine Company
Ziretta
Chianti
Glass, printed paper
Vintage Italian Chianti

14.
Jackson Russell
Jack Russell
Syrah
Glass, printed paper
Vintage Syrah

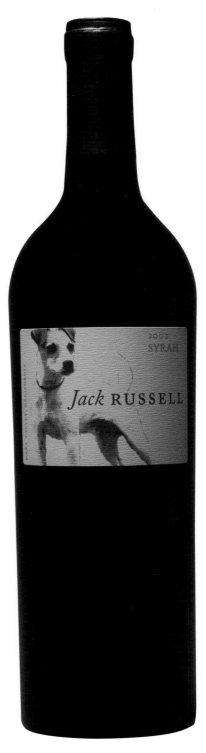

KHDESIGN GMBH
GERMANY

ACTIVE BRANDS
Dynamic markets demand strong brands. But these same brands must remain flexible. A lasting presence in the market demands the ability to adapt. Flexible brand evolution strategies enhance product loyalty and capitalise on previous user experience. A successful brand is a living orga-nism, interacting with its environment in myriad ways: this activity is the source of the brand's vitality in the marketplace.

MOBILE CONSUMERS
Society today is more mobile than ever before. This is true not just in a physical, geographical sense: it applies equally to forms of expression, changing fields of interest and responsibility; to today's dynamic lifestyles. This mobility is re-flec-ted in the choices people make, in their iden-ti-fi-cation with products. Trend controlling is one of the methods we integrate in any brand management strategy, in order to predict change and offer pro-active guidance.

HOLISTIC VISION
The basis for evolutionary brand management is an integrated system, encompassing all the ele-ments which together project the image and quali-ties of a brand. Core values must remain identical, irrespective of whether the brand is communi-cat-ing via POS or a complex multimedia presen-ta-tion. Only a brand architecture that meets these challenges can go on to adapt to changing market requirements and truly add value to a product.

khdesign gmbh
Lilistraße 83 D/09, D-63067 Offenbach, Germany
Phone: +49 (0)69 / 97 08 05-0, Fax: +49 (0)69 / 7 07 83 71
E-Mail: info@khdesign.de, Internet: www.khdesign.de

MEMBER OF
GLOBAL
DESIGN
SOURCE
EXPERTS IN
BRANDING &
PACKAGING
www.g-d-s.net

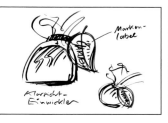   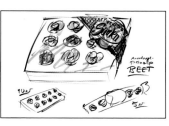 

1. Ferrero
2. Ferrero Garden
3. Non-chocolate confectionary
4. Various materials (paper, foil, aluminum coated paper board)
5. Confectionary for the summer season

1. Pfungstädter
2. Pfungstädter Lemon 1:1
3. Lemon beer mix
4. Aluminum coated paper
5. Young and fresh summer beverage

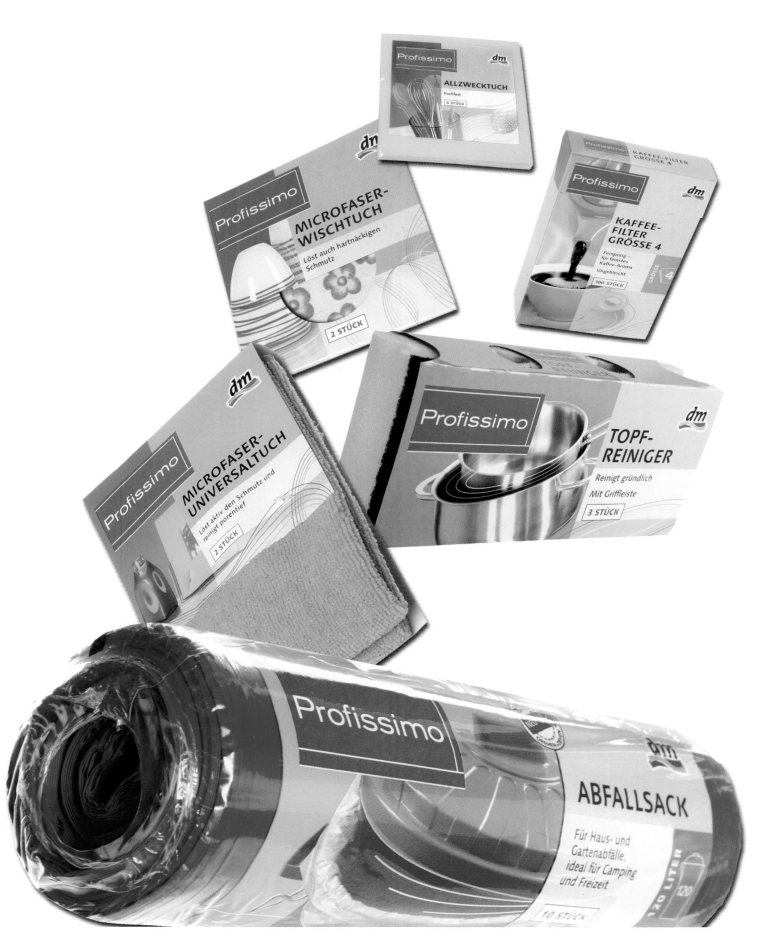

1. dm-drogerie markt
2. Profissimo
3. Household products
4. Various materials (paper, foil, aluminum coated paper board, etc.)
5. New appearance for various household products

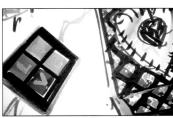

1. W. F. Kaiser
2. Love Collection
3. Baking dish
4. Coated paper board
5. New appearance of modern baking dishes

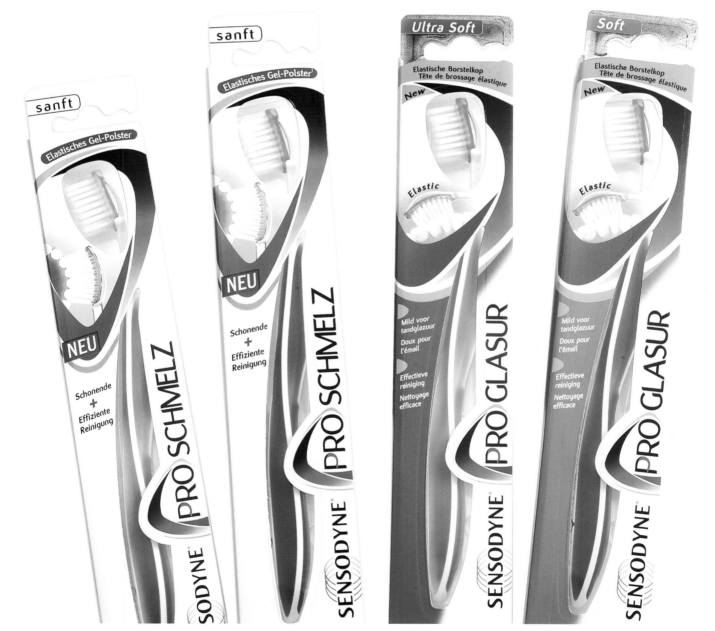

1. GlaxoSmithKline
2. Sensodyne
3. Toothbrushes
4. Coated paper board
5. New medical and modern appearance of high-quality toothbrushes

BRANDNEW·DESIGN

THE NETHERLANDS

Brandnew·design is different, Brandnew·design has Stand Out.
We aim to challenge and to change. We have chosen for a single minded focus on Branding, Packaging and Structural design.
We believe the creation of Stand Out is the fundamental basis for Brands to perform. However no Stand Out without a clear idea of what the Brand Stands For. For us and for Brand owners. Strategy and ambition are combined in creating a visual idea that expresses relevancy to consumers.

That's who we are, that's what we do, that's Simply Stand Out.

www.brandnew.nl

Beauty 1
KAO Brands Europe
Guhl Hair coloring

BRANDNEW·DESIGN

Leeuwenveldseweg 18 1382 LX Weesp Netherlands Phone: +31 294 492 149

Food   2
Lambweston
Twisters Supreme potatos

Food 3
Albert Heijn
Excellent frozen oriental snacks

Food 4
Go Tan
Meal kit

Food 5
Friesland Foods
Milk & Fruit drink

BRANDNEW·DESIGN
Leeuwenveldseweg 18 1382 LX Weesp Netherlands Phone: +31 294 492 149

Drinks 6
Koningshoeven
Speciality beer
brewed by Monks

Food 7
Maggi
Seaesoning for everyday
meat variation

Food 8
Sultana
Healthy in between
meal snack to stay in shape

Food 9
Frico
Gouda de Mai

Beauty 10
Kneipp
Bath oils

Agency promotion 11
Brandnew-design
Out of the Box project
Forgotten Vegetables

THE BARRIE TUCKER COMPANY
AUSTRALIA

Barrie Tucker
Designer/Design Consultant
The Barrie Tucker Company P/L
PO Box 390 Nairne 5252
South Australia
Phone: +61 8 8388 0236
Fax: +61 8 8388 0239
Mobile: +61 (0)439 994 000
barrie@barrietucker.com
www.barrietucker.com

Barrie Tucker bottle design in USA
Represented by Erica Hiller Harrop
Global Package LLC
PO Box 634 Napa
CA 94559 USA
Phone: +1 707 224 5670
Fax: +1 707 224 5683
Mobile: +1 707 294 7899
eharrop@globalpackage.net
www.globalpackage.net

Barrie Tucker AGI LFDIA
Alliance Graphique Internationale
Life Fellow Design Institute of Australia

Barrie Tucker is one of the world's most respected wine packaging and bottle designers.

For over 30 years, Barrie's exciting design creations have won many high profile international awards and have provided outstanding sales results for his clients.

1.
1. Mission Hill Family Estate/Fork in the Road Vineyards (Canada)
2. 'Fork in the Road'
3. Red and White Wine
4. Glass Bottles, Pressure Sensitive Labels, Printed Screw Capsules
5. Brand identity, label and capsule designs by Barrie Tucker

2.   1. ARH Wine Company (Australia/Export to USA)
     2. 'Clarendon'
     3. Premium Red Wine
     4. Exclusively made glass bottle, identity cartouche embossed in shoulder and sand blasted. Pressure sensitive labels. Printed embossed capsule
     5. Identity, bottle and label design by Barrie Tucker

3.   1. Burton Wine Company (Australia)
     2. 'Burton'
     3. Premium Red Wine
     4. Exclusively made glass bottle, identity cartouche embossed in shoulder. Printed capsule. Paper label.
     5. Identity, bottle and label design by Barrie Tucker

4. 1. Yalumba Wine Company (Australia/Export to UK)
    2. 'Yalumba Organic'
    3. Premium Organic Red Wine
    4. Glass bottle, pressure sensitive paper labels
    5. Identity and label design by Barrie Tucker.
       Calligraphy by Jody Tucker (Tucker Creative)

5. 1. Yalumba Wine Company (Australia/Export to UK)
    2. 'Yalumba Organic'
    3. Premium Organic Red and White Wine
    4. Glass bottle, pressure sensitive paper/wrap around labels, printed screw caps
    5. Identity and packaging design by Barrie Tucker.
       Calligraphy by Jody Tucker (Tucker Creative)

6.
1. Chain of Ponds Wine Company (Australia)
2. 'The Ledge'
3. Premium Red Wine
4. Glass bottle, pressure sensitive labels
5. Identity design and label design by Barrie Tucker

7.
1. Chain of Ponds Wine Company (Australia)
2. 'The Amadeus'
3. Premium Red Wine
4. Glass bottle, pressure sensitive labels
5. Identity design and label design by Barrie Tucker

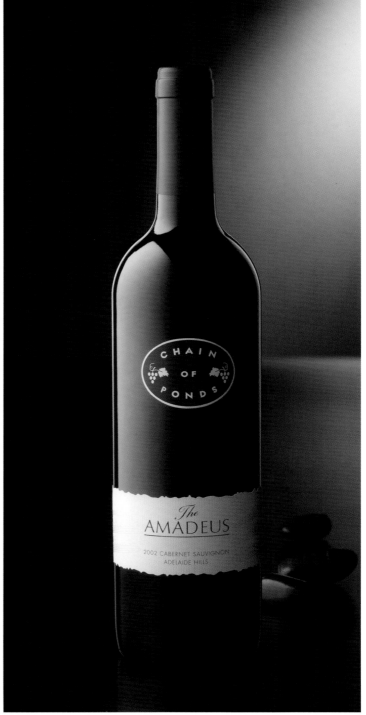

8.
1. Chain of Ponds Wine Company (Australia)
2. 'Nebiolo, Pinot Grigio, Sangiovese'
3. Italian Style Wines
4. Glass bottle, screw caps, pressure sensitive labels
5. The three Italians concept, identity design and label presentation design by Barrie Tucker

9.
1. Chain of Ponds Wine Company (Australia)
2. 'Kangaroo Island'
3. Red Wine
4. Glass bottle, pressure sensitive wrap around paper labels
5. Illustration from old engraving. Design of labels and identity by Barrie Tucker

10.
1. Penley Estate (Australia)
2. 'Penley Pinot'
3. Sparkling Wine
4. Glass bottle, pressure sensitive label
5. Design by Barrie Tucker. Painting by Barbara Chalk

11.
1. Australia Post, Philatelic Group
2. 'Fly Away Flett'
3. Apple Schnapps
4. Mouth blown glass bottle, slumped glass and copper wing. Paper label
5. One of two bottles made as farewell gifts to two members of Stamp Advisory Committee of Australia Post. Bottle mouth blown by Nick Mount to a concept drawing by Barrie Tucker

12.
1. Barrie Tucker
2. 'Purple Panache'
3. Concept bottle
4. Mouth blown glass bottle, silver leaf, Lightning Ridge opal set in silver
5. Concept bottle design by Barrie Tucker.
   Mouth blown (one-off) bottle by Nick Mount.
   Silverwork by Peter Coombs

TUCKER CREATIVE
AUSTRALIA

Tucker Creative - achieving excellence in packaging for over 30 years.

Tucker Creative is one of Australia's leading packaging and branding specialists. Formerly known as Tucker Design, the name change was necessary to better describe the complete range of creative services we now offer our clients. Excelling in all disciplines of Design, Marketing and Advertising, Tucker Creative is a full service creative studio with a reputation for producing innovative, outstanding and inspiring work.

Due to our complete creative service offerings, the Tucker Creative advantage is really our ability to take any brand from conception – right through to enjoying resounding market success. And this success is by no means limited to Australia. We have vast experience in strategically developing new brands, or re-positioning existing ones, to command their price points, stand out on retail shelves and become long-term leaders in the United States – and many other global markets.

Recently celebrating our 30th year in business, Tucker Creative has accumulated a huge folio of work, a cupboard full of prestigious international awards and a proven track record for providing successful and compelling packaging and branding solutions.

Over the next eight pages, we hope you enjoy a cross section of our more recent work. For further examples, or for more information, please call +61 8 331 1700 or visit www.tuckercreative.com.au

Head Office & Studios
57c Kensington Road
Norwood South Australia 5067

T +61 8 8331 1700
F +61 8 8331 1222

USA
Napa Valley
P.O. Box 634
Napa, CA 94559

T +1 707 224 5670
F +1 707 224 5683

Bay Area
3243 Wyman Street,
Oakland, CA 94619
T +1 510 532 6563

tuckers@tuckercreative.com.au
www.tuckercreative.com.au

1
1. Oliver's Taranga Vineyards, Australia
2. Revolution
3. Wine
4. Glass bottle, ceramic print, natural cork, PVC capsule
5. Bold new direction for next generation (from traditional wine making family). Premium wine, US market.

2
1. Samuel's Gorge, Australia
2. Samuel's Gorge
3. Wine
4. Glass bottle, self-adhesive label, natural cork, PVC capsule
5. Contemporary mosaic representation of regional landscape. Super-Premium wine, US & Australian markets.

3
1. Shelmerdine Vineyards, Australia
2. PHI
3. Wine
4. Glass bottle, self-adhesive label, pewter brand disc, natural cork, PVC capsule
5. A true reflection of a unique union of family and vineyard. Limited edition ultra-premium wine, Australian & UK markets.

4 1 Leasingham Wines (Hardy Wine Company), Australia
   2 Leasingham
   3 Wine
   4 Glass bottle, self-adhesive label, natural cork, PVC capsule and screwcap closures
   5 Regional and historic wine range reviewed for stronger brand family image, global market.

5　1　Barossa Valley Estate / Hardy Wine Company, Australia
　　2　Barossa Valley Estate
　　3　Wine
　　4　Glass bottle, self-adhesive label, natural cork, PVC capsule and screwcap closures
　　5　A more contemporary and progressive packaging solution for a Premium to Icon wine in a global market.

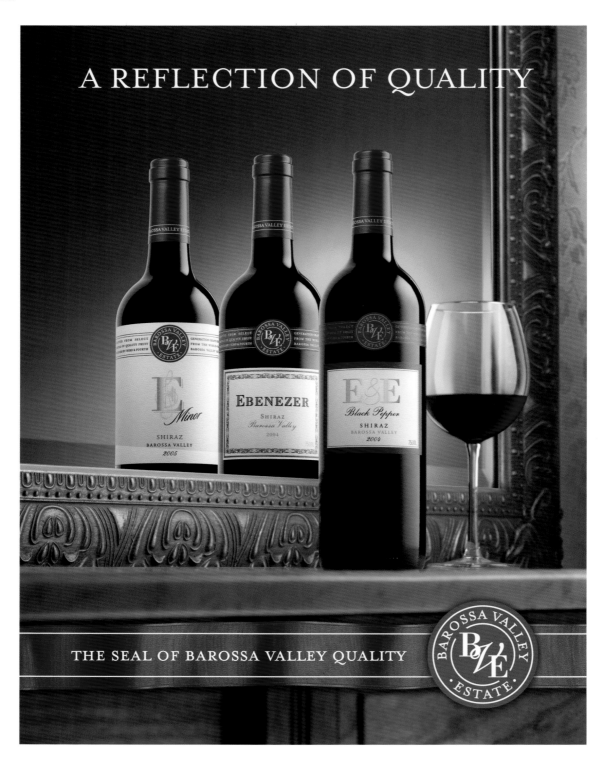

6 1 Sirromet Wines, Australia
    2 820 Above
    3 Wine
    4 Self adhesive label, Glass and Screwcap closure
    5 High altitude premium wine for the Australian market.

7 1 Sirromet Wines, Australia
  2 Australian Sunshine, Australian Sunset, Australian Sparkling Red
  3 Wine
  4 Glass bottle, self-adhesive label, screwcap closure
  5 Exclusive Duty free product catering to Asian tastes with an iconic Aussie look.

8 1 Sirromet Wines, Australia
  2 Sirromet
  3 Wine carry packs
  4 Cardboard
  5 Memorable take away wine carry pack.

9 1 Powercell (Australia) Trading Pty Ltd, Australia
2 Powercell
3 Primary and Secondary Dry Cell Batteries
4 Dry Cell Batteries, Card & clam shell
5 Specialist cell products servicing niche Australian markets.

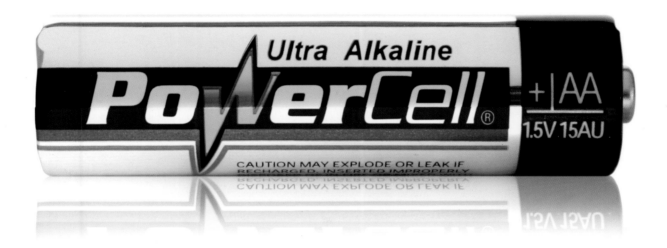

the Lemon Yellow Design
TAIWAN

In 1979, Lemon Yellow stood out as the first and only design house in Taiwan. It has been the design company's goal and duty to provide clients with "tailor-made" service including product introduction, sales promotion, PR events, corporate identity system (CIS) and intergrated communication.

We and our clients have grown together by facing changes and challenges side by side for years. As a forerunner specializing in visual design, Lemon Yellow has helped to crate a new era for the industry by promoting the right image of the manufacturer's products to consumers.

BETC DESIGN

FRANCE

WHO IS BETC DESIGN ?
BETC Design Group is the design division of Havas Advertising, the 5th largest Communications & Marketing Group Worldwide. We have the benefit from the client Intelligence Departement of Havas Group.

OUR PHILOSOPHY & MISSION
Designing for people as a "Creative Factory" in a spirit of a holistic design approach is our mission. BETC Design Group commitment is to answer our clients on several key-ideas : Corporate branding, Product innovation, Retail store design (merchandising).
As a matter of fact, our creative workshop focuses on both symbolic & concrete relationship between Brand, Product and Consumer. As design process aims at generating values & meaning, the originality of our method stems from our user-centered approach : an understanding of individual's behaviours and motivations. Consequently our design process considers both experience of purchasing & product use.

OUR ADDED VALUE
Anticipation : consumer behaviours & expectations, styles & trends
Creativity : Strategic, Conceptual & Technological
Management : design process, brand identities and new product lines

BEST CASES : 2 EXAMPLES
In the transport field we have designed 2 major sets :

- The Bordeaux Tramway, a safe and comfortable transportation environment, designed and realized with a strong partnership between our in-house engineers and ALSTOM engineers. Awarded the "Strategy Product Design Prize" in 2003.
- The AIR FRANCE new First Class and Business Class, from an in-depth study of passenger expectations we have conceived seats as a tangible and constant extension of on-board service and the care given to the passenger.

www.betcdesigngroup.com

Amore Pacific
Laneige cosmetic brand
Range of take away packaging system

IFRI
Algerian Softdrink
New range of labels for 6 flavors

Canal +
Decoding machine
New range of packaging system

GEMOLOGY®

Gemology
French cosmetic brand for mature skins
Brand Identity, and new range of cosmetic packaging development

Amore Pacific
Laneige cosmetic brand
Range of gift packaging system

Charles & Piper Heidsieck Champagne (Rémy Cointreau Group)
Charles Heidsieck Champagne
Range of 3 relifted packagings

Charles & Piper Heidsieck Champagne (Rémy Cointreau Group)
Champagne bucket, limited edition 2006-2007
Product design in collaboration with Jaime Hayon

STRØMME THRONDSEN DESIGN
NORWAY

The designers Dagheid Strømme and Morten Throndsen set up Strømme Throndsen Design in 1995. Strømme Throndsen design is now one of Norway's leading design agencies, with considerable experience in concept development and graphic design within the area of packaging identities, corporate identities and 3D design.

The key to our existence is to strengthen the competitive advantage of the company or the brand by developing unique and high quality design concepts. Design plays an important strategic role in building brands. Our philosophy is to build brands based on strategic thinking and create stories that help the strategy come alive in an innovative and distinctive way. Concept development is therefore at the heart of our creative process.

Strømme Throndsen design serves many of Norway's leading FMCG companies and has proven itself as a successful design agency over 10 years. The award-winning agency has been nominated for the prestigious Honours Award for Design Excellence by the Norwegian Design Council 3 times, and has received the Award for Design Excellence a numerous times.

1. Client: Løvenskiold / Country: Norway / Brand: Løvenskiold / Type of product: Filet of moose / Material: Cardboard / Theme: Severin Løvenskiold - Norwegian prime minister 1828 - 1841

2. Client: Lillehammer Ysteri / Country: Norway / Brand: Lyst! / Type of product: Range of cheese / Material: Environmentally friendly plastic / Theme: Appetite! (Lyst!) for cheese
3. Client: Rieber & Søn / Country: Norway / Brand: Black Boy / Type of product: Range of Spices / Material: Tin can / Theme: Kitchen style

4. Client: Steen & Strøm Magasin / Country: Norway / Brand: Steen & Strøm Magasin / Type of product: Shopping bag / Material: Paper / Theme: Norway's oldest and most prestigious department store

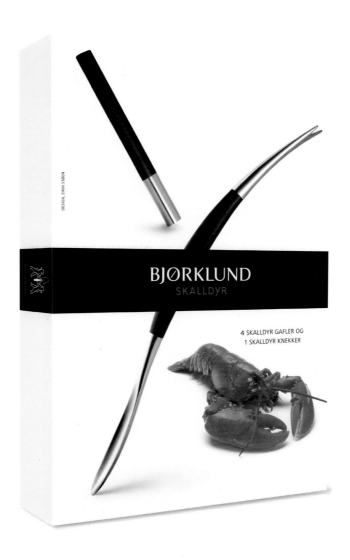

5. Client: Bjørklund / Country: Norway / Brand: Bjørklund Skalldyr / Type of product: Range of seafood cutlery / Material: Cardboard / Theme: Pure Norwegian

6. Client: Rignes / Country: Norway / Brand: Rignes Swing / Type of product: Beer / Material: Glass / Theme: Loosen up!

MOUNTAIN DESIGN
THE NETHERLANDS

Mission Statement

People are looking for brands that express 'something bigger than themselves'. Brands with a story. Strong, ambitious brands that represent a clear vision, a committed mission and a distinctive culture. We want to dedicate our creativity to realising and delivering those inspiring brands.

We are an international design agency with a dedicated team of creatives which started life in 1998. Originally focused on innovative and distinctive packaging, our passion now extends to brand strategy and corporate identity. We are passionate thinkers and creative doers. We believe in starting from a solid strategic ground, then building an inspired 360 degrees around it.

www.mountaindesign.nl

Campina Fresh dairy products

Groene Koe Biological dairy products

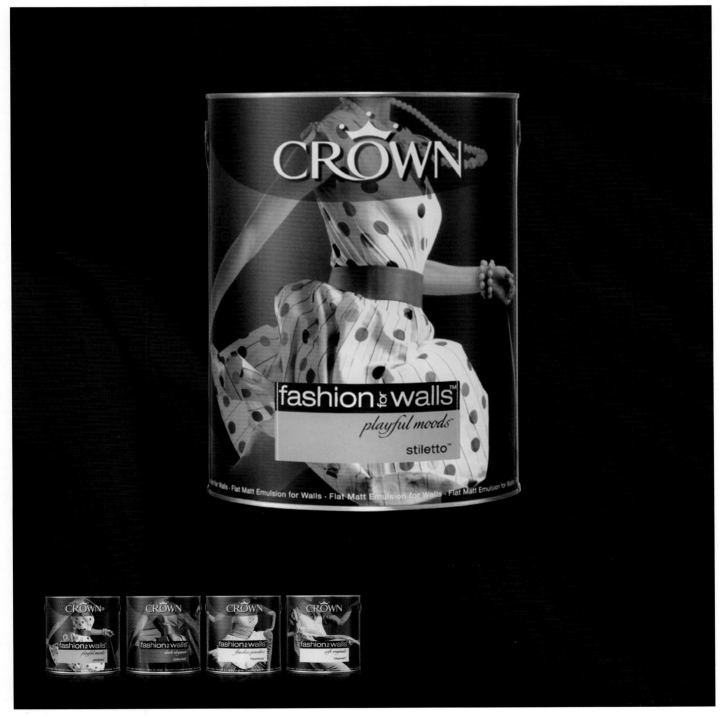

AKZO Nobel, Crown Fashion for Walls paint

Coca Cola Germany, Appolinaris waters

Leaf, Truly range

Inbev, Hertog Jan beers

Coca Cola Netherlands, Fanta range

Dr vd Hoog cosmetics, Pure Line

DEUTSCH DESIGN WORKS

U.S.A.

In 1995, Barry Deutsch and Lori Wynn founded Deutsch Design Works (DDW) with the idea of combining creativity with strategic thinking to generate ideas that engage and enlighten. More than a decade later, toting multiple awards and a storehouse of global brand expertise, DDW brings inventive, on-target dedication to its branding, packaging, and identity design.

Located in San Francisco's Potrero Hill district, DDW's unusual courtyard environment provides an open-campus oasis conducive to letting the creative juices flow. Putting a premium on the imaginative and the ingenious, only the most creative come on board. The voice of experience spiked with fresh energy characterizes the collaboration of DDW's multi-national, talented design team.

Since inception, DDW's lunchbox logo has symbolized the company's roots... pack a lunch, pull up a chair – explore possibilities.

10 Arkansas Street
San Francisco, CA 94107
415.487.8520
www.ddw.com

Albertson's Market: Essensia Brand Fresh Fruit Pies

Attune Wellness Bars
A naturally flavored snack that gives your body energy and the benefits of yogurt.

Dreamerz
A natural beverage that helps induce a restful night of sleep.

Joint Juice
A fruit flavored drink that supplies one with a daily dose of glucosamine that relieves joint pain.

Sinvino
A sophisticated sparkling natural beverage, lightly carbonated with apple and black currant flavors. Bottle and branding design.

Shakers Vodka
An American vodka made from wheat. We designed the bottle to reference a martini shaker that came of age in the 1930's – graphics followed.

Solaris Wines
An offering from Diageo to compete in the medium price category. We created an icon using sun flare photography from NASA.

Precis Vodka
A Swedish import bottled in hand-blown glass and silver closure. The name is derived from the "precision" distillation process.

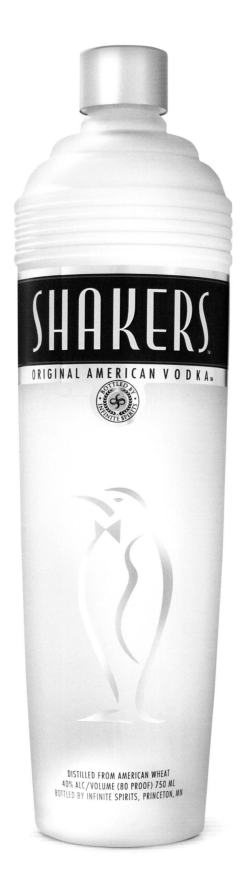

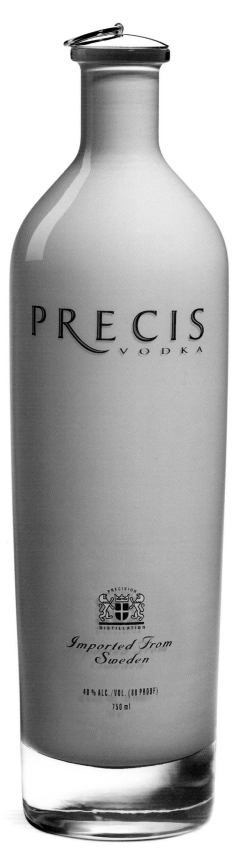

Bellagio Casino & Resort
The logo design we created to echo the style of the Italian lakeside town and luxury of Las Vegas.

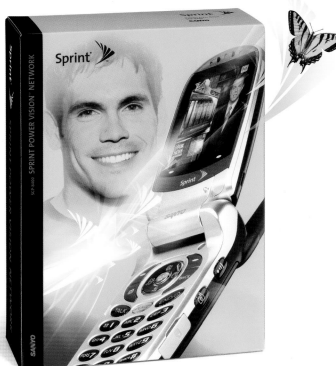

Sprint Wireless Phones
A line of boxes featuring the latest and most sophisticated technology.

Mojito
A hip, mint flavored malt liquor addition to the Bacardi Silver line of beverages.

Diet Pepsi
We created the color, look and feel for the sub-brand that became Pepsi's flagship beverage.

Hansen's Natural Soda
The colorful cans tell the "flavor" story and the California fruit crate style convey the brand's heritage.

Annie's Homegrown Cereals
The newest product under the popular Annie's brand creates a series of "window icons" for each flavor and displays the fanciful product shapes.

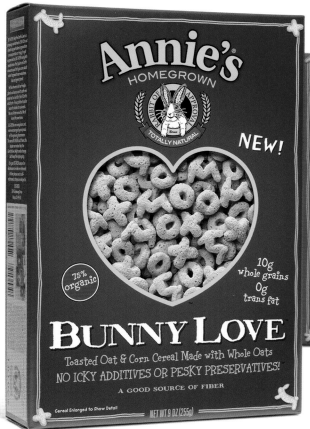

LightFull Satiety Smoothies
A new product for women-on-the-go!
A healthy mid-day snack that's filling and
delicious. Less calories, more filling!

Mug Root Beer
A redesign of Pepsi's
young male beverage –
emphazising the product's
fun and "foamy" character.

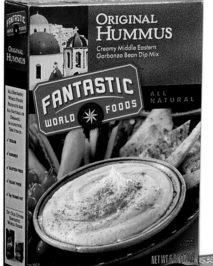

Fantastic World Foods
This recent redesign of Fantastic's Brandmark and Packaging
brought to life the new positioning of "never ending discovery."
Combining stock photos for "place" and delicious food shots,
we created 65 soups, sides and meal boxes for the brand roll-out.

CATO PARTNERS

AUSTRALIA

Australia

Melbourne
Cato Purnell Partners
10 Gipps Street
Collingwood 3066
Victoria
Telephone +61 3 9419 5566
Facsimile +61 3 9419 5166
melbourne@cato.com.au

Sydney
Cato Purnell Partners
Suite 84, The Lower Deck
Jones Bay Wharf
26-32 Pirrama Road
Pyrmont Point 2009
New South Wales
Telephone +61 2 9660 4555
Facsimile +61 2 9660 7766
info@catosydney.com

Adelaide
c/- Einstein Da Vinci
94-96 Fullarton Road
Norwood 5067
South Australia
Telephone +61 8 8362 5211
Facsimile +61 8 8362 8633

Brisbane
84 Cavendish Street
Nundah 4012
Queensland
Telephone +61 7 3314 6229

Perth
c/- Forbes Agency
The Boans Warehouse
10 Saunders Street
East Perth 6004
Western Australia
Telephone +61 8 8362 5211
Facsimile +61 8 8362 8633

International

New Zealand
Cato Partners NZ Limited
Level 9
175 Victoria Street
Wellington
Telephone +64 4 499 5549
Facsimile +64 4 499 6226
mail@cato.co.nz

Singapore
Cato Consulus Pte Ltd
11B Bali Lane
Singapore 189848
Telephone +65 6293 9495
Facsimile +65 6293 9485
lawrence@consulusgroup.com

Spain
Cato Partners Europe
Calle Costa Brava, 13 2
Madrid 28034
Telephone +34 91 735 5689
astalman@catospain.com

United Arab Emirates
Cato Partners
Mot2iwalla
The Complete Digital Resource
Telephone +971 4 359 6188
Facsimile +971 4 355 2889
Dubai 31788
motiw@emirates.net.ae

México
Cato Saca Partners
Terranova 714-A int.2
Providencia
Guadalajara 44670
Telephone +52 33 3640 6566
Facsimile +52 33 3617 1365
rsaca@catosacapartners.com

India
Design Protocol Pvt. Ltd F.
Unit 10, Drego House
Dr Ambedkar Road
Bandra (West)
Mumbai 400-050
Telephone +91 22 2605 1579
Facsimile +91 22 2600 8075
designprotocol@vsnl.net

Ken Cato, Chairman

Graham Purnell, Creative Director

Cato Partners was established in 1970 and has since grown by providing all sectors of Australian and international business with brand strategy and identity management and design.

Cato Partners is now one of the largest identity management and design companies in the Southern Hemisphere and has worked across a broad depth of branding and design projects, in over 32 countries. Cato Partners' experience is second to none, with the team responsible for delivering on dozens of local and international projects each year, within almost every type of category, demography, culture and application.

The nature of the profile of clients has demanded that Cato Partners must simultaneously create work that is visually distinctive, enduring and versatile. Cato Partners have developed some of the most successful brand strategy and identity projects worldwide. The ability to achieve this is based on the outstanding calibre of a highly creative design team, experience with organisations globally for over 37 years and the application of proprietary methodologies.

At Cato Partners it is believed a 'Broader Visual Language' must be established for each organisation or brand. Most importantly, Broader Visual Language™ enables the development of a comprehensive and consistent identity for all sectors of the business. It provides synergy between the master brand and sub brands, establishing strong market presence and making the overall business strategy viable and visible. It requires an outstanding design team to interpret the strategy of Broader Visual Language™ into design language that endures with great visual distinction. At Cato Partners the design team has brought its talents to fruition on major projects worldwide

website www.catopartners.com.

An Australian designer with an international reputation, Ken's prolific work encompasses all facets of corporate and brand management and design. His philosophy of design is dynamically holistic, providing the synergistic solutions that produce positive results. He has won countless international design awards, and his work is represented in museums and galleries around the world.

A long-standing member of Alliance Graphic Internationale and a past AGI President, Ken has previously been awarded the first Australian Honorary Doctorate of Design and inducted into the 'Hall of Fame' of the inaugural Victorian Design Awards. He is also a foundation member of the Australian Writers and Art Directors Association, a member of the American Institute of Graphic Arts, ICOGRADA, Design Institute of Australia, Australian Marketing Institute, Industrial Design Council of Australia, and is Patron of the Australian Academy of Design.

Ken is regularly invited to speak at international symposia and is an acclaimed author who has written numerous books, including First Choice, Design for Business, The View from Australia and Design by Thinking. In 2006 Ken was honoured with the Premiers Award for design leadership from the Premier of Victoria.

Graham's unwavering commitment to excellence in design has led to an enviable international reputation. His considerable talents and expertise can be seen in international design publications including D&AD, Design Downunder and Graphis. His highly developed skills and creative approach have led to the success of thousands of projects for hundreds of clients around the world.

Graham completed a Master of Arts at the Royal College of Arts in London in 1984 before taking up the position of Senior Designer with Minale Tattersfield. Subsequently, Graham spent over three years working in Singapore as Design Director enhancing his skills in the strategic area of corporate identity design, packaging and annual reports.

Since joining Cato Purnell Partners in 1994, Graham was appointed Creative Director in 1998 and became Company Partner with Ken Cato in 2001. He has been involved with a wide variety of projects leading the design teams. His highly developed skills and creative approach have been vital to the success of projects for clients such as Energex, Coles Farmland, Kraft, Nestlé Nescafé, Primelife, Royal Guidedogs Association, Suncorp Metway, Foster's Australia, IAG, Woolworths, Burswood Entertainment Complex, Barry Plant, RedKite, Frito Lay Mexico, Carta Blanca BenQ and many others.

1-2

4

3

1-2   1. BenQ Siemens
13.   Mobile Phone Package
21.   BenQ
23.   CD Package
31.   Norsewear
      3. Footware Packaging
41   Woolworths
      2 Select
      3 Premium Grocery Range

5 1. Tarrawarra Estate
   2 Tin Cows Chardonnay
6 1 Underground
   2 Underground Merlot
7-9 1 Foster's Group Limited
7 2 Light Ice
7 3 Chill Filtered Beer
8 2 Victoria Bitter
8 3 Beer
9 2 Carlton Draught
9 3 Draught Beer

5

6

7

8

9

10   1. Tarrawarra Estate
      2. Tarrawarra Estate Pinot Noir
11   1   Foster's Group Limited
      2   Crown Lager
      3   Premium Beer

12 1. Melba Recordings
    2 Seduction
12 3 Album Covers
13 1 Yoga Away Corporation
    3 DVD Packaging
14 1 AGIdeas
    2 International Design Conference
    3 Program Packaging

12

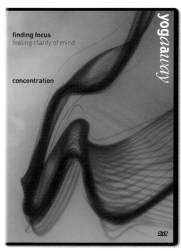
13

14

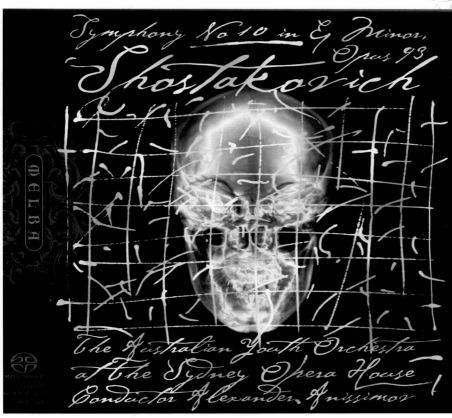

15

16

15   1. Melba Recordings
      2 Shostakovich
      3 Album Covers
16   1. Melba Recordings
      3 CD Packaging

BENEDICT CORPORATION
JAPAN

MASAKAZU TAGAWA

Born 26th July 1955 in Tokyo, Japan.
Graduated Nihon University 1978. Established Benedict corporation in April 1997.
Won Silver Award at the Japan Package Design Award in 1997,
the Special Award at the Japan Package Design Award in 1998,
the sectional award at JPC in 1999, the Japan Star Award at the Japan Package Contest in 2000,
the Ministry of Economics, Trade & Industry's Commissioner's Award at JPC in 2001,
and won at the first Free & Easy Design of the Year Awards in 2007.

Phone:81-3-3492-7401   Fax:81-3-3492-7402
Mail:info@benedict.jp   HP:http://www.benedict.jp

1 Asahi Soft Drinks / Coffee

Bireley's Orange

3 Asahi Soft Drinks / Fruit Drink

4 Tilia / Hand Maid Teddy Bear Gift Box
5 Tilia / Embroidery Broach Gift Box

DIL BRANDS

BRASIL

DIL BRANDS
Corporate and Consumer Branding

Founded in 1961, DIL Brands is dedicated to strategic branding and design for consumer and corporate brands.

We are able to manage entire branding programs, from structured brainstorming sessions, ideation and concept sketching, to product concept (product, graphic and industrial design) and precise, coordinated implementation.

If there is a need for focused and structured creativity, locally of globally, DIL Brands has all the tools and expertise to do it.

- More than 5,000 new packages over 45 years
- 4 live-connected offices: Brasil – Chile – Argentina – México.
- Multicultural, multiracial, multimotivated
- In-house prototyping capabilities
- Videoconference capabilities
- The most awarded consultancy in our region - Worldstars, Clios, London Awards;
- A CBX partner

Rather than dedicating ourselves to creating simply beautiful and attractive designs, we are focused on delivering "designs that sell". This is the intriguing nature of our business and the essence of our quality standards.

CBX (www.cbx.com) is our worldwide resource, a network of world-class brand strategy and design consultancies that stretch our reach to Europe (UK and The Nederlands), United States (New York, Minneapolis and San Francisco), Oceania (Melbourne) and Asia (Ghengzhou and Shangai – China). Whatever the need, we will speak the language.

SAO Al. Rio Negro 1030 - cj. 2304 - 06454-000 Alphaville SP, Brasil (55 11 4191 9711)
MEX Jaime Balmes 11 Piso 7 Polanco Chapultepec, México (52 55 5580 4047)
SCL Andrés Bello 2777 – of. 2403 Las Condes Santiago, Chile (56 2 203 3844)
BUE Sinclair 2949 Piso 8 C 1425BXO Buenos Aires, Argentina (54 11 4139 5485)
www.dilbrands.com

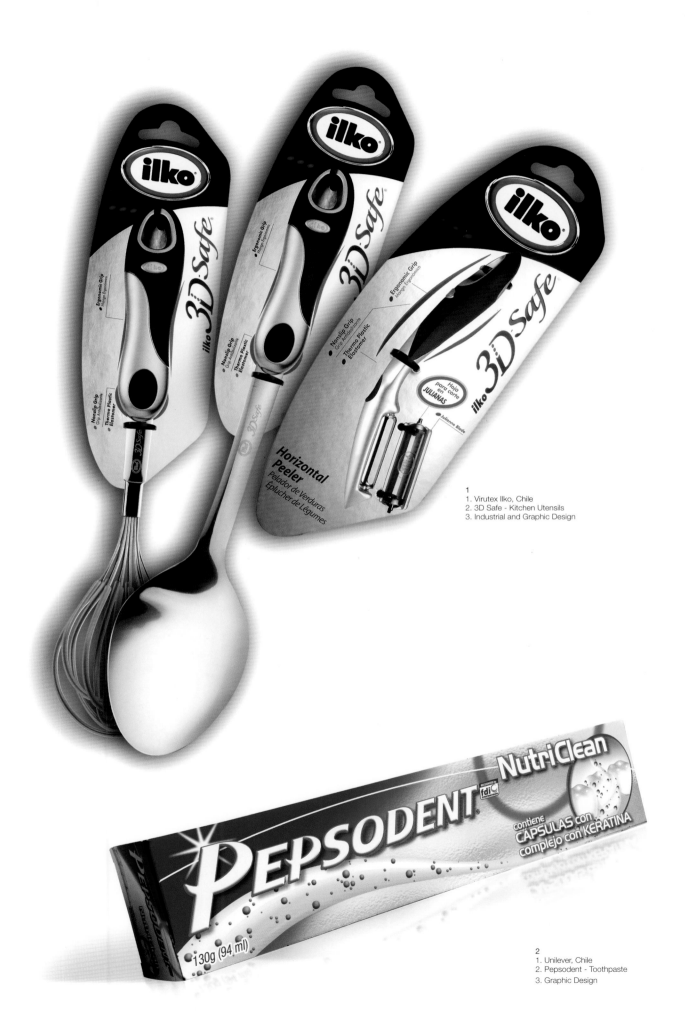

1
1. Virutex Ilko, Chile
2. 3D Safe - Kitchen Utensils
3. Industrial and Graphic Design

2
1. Unilever, Chile
2. Pepsodent - Toothpaste
3. Graphic Design

3
1. Unilever Foods, Latin America
2. Hellmann's Ketchup
3. Industrial and Graphic Design

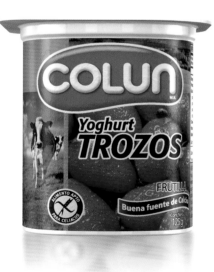

4
1. Colún, Chile
2. Colún Brand Identity – Dairy Products
3. Graphic Design

5
1. Capel, Chile
2. Maddero Rhum
3. Industrial and Graphic Design

6
1. Viña Santa Rita, Chile
2. 120 3 Medallas Wine
3. Graphic Design

7
1. Nestlé
2. Nestlé Chocolate Drink
3. Graphic Design

8
1. Molinos, Argentina
2. Matarazzo Premium Pasta
3. Graphic Design

9
1. Nestlé
2. Nescafé
3. Graphic Design

10
1. Nestlé, Argentina
2. Heaven Premium Ice Cream
3. Graphic Design

11
1. Campbell Soup Co., México
2. V8 Splash Fruit Juices
3. Graphic Design

12
1. Campbell Soup Co., México
2. Campbell's Homemade Soups
3. Graphic Design

13
1. Coca-Cola, México
2. Minute Maid Orange Juice
3. Graphic Design

14
1. Jerónimo Martins, Portugal
2. Pingo Doce
3. Graphic Design

15
1. Jerónimo Martins, Portugal
2. Pingo Doce
3. Graphic Design

16
1. Reckitt Benckiser, Brasil
2. Veet Depilation System
3. Industrial and Graphic Design

17
1. DM Monange, Brasil
2. Supra Sumo Sore Throat tablets
3. Graphic Design

18
1. FEMSA, Brasil
2. Sol Beer
3. Panamerican Games Design

19
1. Reckitt Benckiser, Brasil
2. Veja Surface Cleaner
3. Industrial and Graphic Design

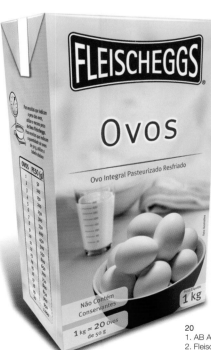  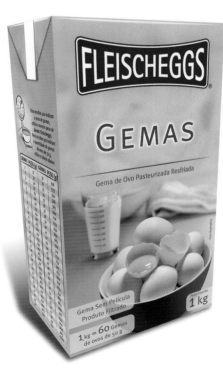

20
1. AB Alimentos, Brasil
2. Fleischeggs Ready to mix eggs/yolks/whites
3. Graphic Design

YASHI OKITA DESIGN
U.S.A.

Yashi Okita Design was founded in San Francisco in 1982. Led by Yashi Okita, he and his team of writers, illustrators and designers serve consumer, high-tech, medical and financial services clients around the world.

"We only feature consumer-oriented package design for this book. When we work on consumer packaging, we really believe we have to do a lot of marketing research. Creating branding is a very complicated and time-consuming process but we really enjoy it because we learn new information and what the current market is like. These days using internet technology, researching information has become much easier. Yashi has mentioned current technology usage in the essay section at the beginning of this book.

Also, technology is making our package design easier and more free because the printing industry has completely changed. We used to be concerned about how many colors we can use and creating a complicated design using lots of hairlines in color and so forth. Nowadays full color printing jobs are even easier than two or three color printing for both designers and printers. The electronic files created with design software have a lot more freedom."

In thirty years, Yashi is modestly proud of a portfolio that shines with award winning designs from leading industry publications and professional organizations, including HOW magazine, Print magazine, PAP, Creativity, print Industry America, Firm Outlook Design, the Murphy's Award and AAF.

1. Pollack & Company, USA
2. Pollack Chili Oil
3. Bottled flavored olive oil
4. Glass bottle with paper label

1. Moonstar Restaurant, USA
2. Original Barbecue Sauce
3. Bottled barbecue sauce
4. Glass bottle with paper label

1. Moonstar Restaurant, USA
2. Almond Cookies
3. Cookies
4. Plastic wrap and paper label

1. Bear Valley Resort, USA
2. Bear Valley Gift Shop
3. Shopping bag and gift box
4. Paper bag and sleeve

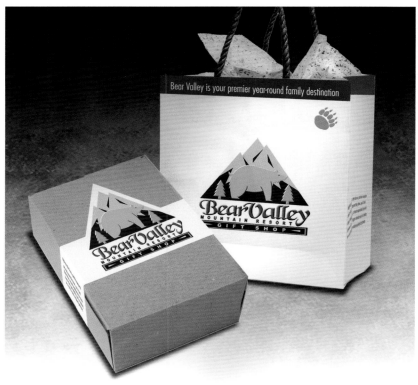

1. Titan International, USA
2. Mishelle Maddox Inspire
3. Women's shoe packaging
4. Paper box

1. Silicon Frontline, USA
2. Silicon Frontline
3. Software packaging
4. Package, CD label and case

UP CREATIVE

TAIWAN

UP CREATIVE Design & Advertising Corporation

4F., No.269, Sec. 5, Zhongxiao E. Rd., Taipei 11065,
Taiwan, R.O.C.
T. +886 2 2765 1181
e–mail:up@upcreate.com.tw
www.upcreate.com.tw

Founded in 1988 and keeping constancy in commercial design, UP CREATIVE is neither a commercial agency nor a fine art creation, but a source of professional suggestion and utility creative design, which awarded by lots of international and local competitions. UP CREATIVE firmly believes that commercial design will eventually go back to trade itself, only under the concept of marketing plans and visual design can provide a useful and unique design to reach the most powerful visual effect in greatest benefit. UP CREATIVE always persists in *"Creation, Quality, Service, and Timing"*. Without a qualified and creative design in time equals to the failure in effective service to clients. Her philosophy —progress and creation— are continuous, because she exactly knows what commercial design demands. Following the footstep of new age, she will keep her design in Utmost Progress.

1   1. Heme International Inc.
    2. heme
    3. Skin care series
    4. AS, MS

2   1. Heme International Inc.   3   1. Heme International Inc.
    2. heme   2. heme
    3. Skin care series   3. Skin care series
    4. AS, MS   4. AS, MS, PVC

4   1. Taiyen Biotech Co., Ltd.
    2. La Meilleure Eau de TAiwan
    3. Distilled Deep Sea Water
    4. Glass, MS

5   1. Unilever Taiwan Ltd.
    2. Dove
    3. Cream shower
    4. PP

6   1. Chou Chin Industrial Co., Ltd.
    2. Natural Nuts
    3. Biscuit
    4. Paper

7   1. Uni–president Enterprises Corp.
    2. Milk Flan
    3. Milk Flan
    4. PP, Paper

8   1. Uni–president Enterprises Corp.
    2. One More Cup
    3. Instant Noodles
    4. Paper

9 1. Hunya Foods Co., Ltd.
   2. Le Papillon
   3. Wedding Cake
   4. PVC, Paper

10 1. Pepsico Foods Taiwan Co., Ltd.
    2. Lays
    3. Potato Chips
    4. Aluminum Pouch, Paper Box

11 1. Hunya Foods Co., Ltd.
    2. Mojo
    3. Wedding Cake
    4. PS, PP

12

13

14

12   1. Fong Nien Fong Ho Enterprises Corp.
      2. Le tea
      3. Fruit tea
      4. PET

13   1. Uni-president Enterprises Corp.
      2. Calpis
      3. Light Yogurt Drink
      4. TetraPak

14   1. Jin Mai Lang Food Corporation
      2. Jin Mai Lang
      3. Green Tea
      4. PET

15
1. Fong Nien Fong Ho Enterprises Corp.
2. Le Power
3. Energy Drink
4. Aluminum Can

16
1. Uni–president Enterprises Corp.
2. Calpis (Day of Beauty)
3. Light Yogurt Drink
4. TetraPak

17
1. Taiwan Gimsun Yan Enterprise Co., Ltd.
2. Dr. Neo
3. Healthy Supplement
4. Paper Box

STUDIO GT&P
ITALY

1

2

1   Diva International, Italy

Borgovivo

Extra virgin olive oil and dressings; a bottle contains 13 ml of product, mainly used for airlines in-flight service

2   Diva International, Italy

Borgovivo

Extra virgin olive oil and Italian dressing

Printing Process: 4-colours offset printing.

3   Azienda Agraria Clarici, Italy

Olio Clarici

Extra virgin olive oil

Printing Process: 4-colours offset printing.

1

2

3

1   Penta Star, Italy

    Adamantis

    A special Italian "grappa" housed in a blown glass bottle

    Printing Process: 2-colours offset printing (blue and silver), hot foil stamping (silver) and U.V. silk screen varnish for the background texture.

2   Diva International, Italy

    Buona Vista

    Lens cleaning papers

    Printing Process: 5-colours offset printing.

3   Diva International, Italy

    Diva International

    Refreshing towels mailing packaging, The box is handmade paper covered

4   Diva International, Italy

    Fria

    Special paper and cotton moist towelettes created for selected perfumeries and chemist's

    Printing Process: 2-colours offset printing, blind embossing and powdered gold for the box; 3-colours gravure printing for the sachet.

4

1   Azienda Agraria Tocchi, Italy
    Cantina Poggio Turri
    Wine packaging
    Printing Process: 2-colours offset printing and dull varnish.

2   Fornace Laterizi Clarici, Italy
    Clarici
    Handmade bricks mailing. The box contains the brochures and a handmade brick sample.
    Printing Process: 4-colours offset printing and dull U.V. varnish.

1   Azienda Agraria Tocchi, Italy
    Cantina Poggio Turri
    Wines (Sagrantino & Grechetto)
    Printing Process: 4-colours offset printing, hot foil stamping (gold)

2   Epicure Garden, USA
    Exir
    Powder, stigma and tablet saffron
    Printing Process: 4-colours offset printing.

3   Cantina Fratelli Pardi, Italy
    Cantina Fratelli Pardi
    Wine (Sagrantino)
    Printing Process: 2-colours offset printing, hot foil stamping (gold)

4   Cantina Fratelli Pardi, Italy
    Cantina Fratelli Pardi
    Wines (Montefalco Rosso 1 Colle di Giove)
    Printing Process: 4-colours offset printing, hot foil stamping (gold)

1

2

3

4

IFF COMPANY INC.
JAPAN

For many people, packages remain agood menory,

especially when they are closely related totheir live.

It's not always possible, though,

to intentionally create package, no matter how hard you try.

this is because the essence of a good design is decided by the era,

rather than the times, people in later generation may likely say,

"Oh that. Yeah, I remember when Iwas just akid,.........."

Handkerchief

TDK / Mini Disc

TDK / DVD-R

TDK / CD-R

125

Authorized by
Coca-Cola Japan Company limited

Coca-Cola Japan / Aquarius

Coca-Cola Japan / Daizunosusume

Coca-Cola Japan / Georgia Emblem Black

Coca-Cola Japan / Georgia Ice Café Vanilla

Coca-Cola Japan / Georgia Five Blend

Coca-Cola Japan / Georgia Mocha Au Lait Hazelnut

Coca-Cola Japan / Georgia Mild Milk Café

Coca-Cola Japan / Georgia Walatte

Coca-Cola Japan / Georgia Mocha Kilimanjaro

Coca-Cola Japan / Georgia Clear Bitter

Coca-Cola Japan / Georgia White chocolat

Mascot / Beens Curry

Mascot / one Chicken Bouillon

Mascot / Yellow Curry

Mascot / Red Curry

Mascot / Paella

Mascot / Keema Curry

Mascot / one Consommé

Mascot / Green Curry

Mascot / Tomato Pilaf

Suntory / Magunum Dry

Suntory / Super Chu-Hi

TRIDIMAGE

ARGENTINA

Front, left to right: Virginia Gines, principal and design director; Hernán Braberman principal and design director; Adriana Cortese, principal and chief creative officer.

tridimage
3D PACKAGING IMAGE DESIGN

Tridimage is an integrated graphic and structural packaging design consultancy, based in Buenos Aires, Argentina. We have been producing creative and distinctive branding solutions for clients worldwide since 1995.

We possess a blend of strategic thinking, boundless 3D creativity, commercial savvy, and flawless execution. We have underpinned our business on the ability to drive value back to our clients through brand-focused, commercially effective solutions.

Our extensive experience in 2D and 3D packaging spans across all industries, including food and beverage, health and beauty, technology and durable goods.

Our packaging designs are backed up by our sound knowledge of manufacturing processes, materials and cost constraints, giving our clients a point of difference in an evermore competitive marketplace. At Tridimage we believe that successful packaging reflects the brand positioning, stands out, has conviction and is cost effective in its application.

We run all of our design projects, many of them for clients around the world, from our Buenos Aires studio. Today's technology allows the communication of ideas, creative work and instant access to local market intelligence necessary for the smooth running of a design project from global to local.

www.tridimage.com

---

tridimage
Av. Congreso 4607
C1431AAB | Buenos Aires
Argentina
T/F+54 11 4542 8982
design@tridimage.com
www.tridimage.com

1 Prodea | Argentina
Cunnington Exclusive
Cola sodas
PET bottles / OPP labels
Structural packaging & label design

2 J. Llorente | Argentina
Extra Toso
Sparkling wine
Glass bottle / OPP label
Brand identity & label design

3 Traceli | Argentina
   Natulen
   Gluten-free cookies
   Cardboard cartons
   Brand identity & graphic design

4 Hawk Company Europe | Germany
   FlashBox
   Car & truck accesories
   Cardboard cartons
   Brand identity & graphic design

5 The Clorox Company | USA
   Clorox Anywhere
   Hard surface daily sanitizing spray
   HDPE bottle
   Structural packaging

7 Grupo Berro/Grupo Osborne | Spain
Magno Solera Reserva
Brandy de Jerez
Glass bottle / Paper label
Structural packaging by Tridimage
Label design by Grupo Berro

8   J. Llorente | Argentina
    Berry & Lemon Champ
    Sparkling wines
    Glass bottle / Paper labels
    Brand identity & label design

9   AGLH | Argentina
    Estancia Las Quinas
    Premium honeys
    Glass jars / OPP label
    Brand identity & label design

10 Mexicana de Bebidas | Mexico
    Frizzante
    Flavored sparkling waters
    PET bottles / OPP labels
    Structural packaging & brand identity

11 Prodea | Argentina
    Cellier
    Mineral waters
    PET bottles / OPP labels
    Structural packaging & label design

12 Haraszthy Vallejo | Hungary
Quixotic
Premium wines
Glass bottles / OPP labels
Brand identity & label design

KAZUMI KAGAWA DESIGN OFFICE

JAPAN

5-11-501-1503 Koyocho-naka, Higashinada-ku Kobe
658-0032 Japan
Telephone : + 81・78・857・1077
Facsimile : + 81・78・846・2312
Mobile : + 81・90・228・49780
E-mail : kagawa@lares.dti.ne.jp
Contact : Kazumi Kagawa

Please note that Kazumi Kagawa Design Office will be moving to Osaka in autumn, 2008.
You can reach me by email or mobile phone from September, 2008 onward.
Please email me if you would like to know the new postal address.

The 21st century —— that is the age in which human communication should be focused on after the age of materialism.
Although the needs of people and the culture they create in a shifting world have been changing all the time, our principal needs must not change.
Human beings can not be human beings without communication.
My office symbol is constructed by my initial K.K., and represents my wish that we will be able to grow up together hand in hand.
I guarantee that I would express "self" concerning both your own requirements and market demand, and would flexibly support your communication business.
I hope it will help you with expanding your business opportunities.

The director of Kazumi Kagawa Design Office,
Kazumi Kagawa is a member of the Japan Graphic Designers Association.
and a part-time lecturer at Kobe Design University.
From 1984 to 1997 she worked at Product planning and development departments, Noevir Co., Ltd. (cosmetics company) in Osaka, Shiga and Kobe, and there designed a variety of products including packages and containers for skin care and hair care products as well as gift items.
She was awarded at the Package Design Council's Gold Award Competition in 1991 for the "Dr.PC-2" and won the R&D Contest Design Award with the "Lily of the Valley (Yume Suzuran) Body Care Series" in 1987.

Design = A high quality interface between consumer goods and the people who use them.
Through my designs, I hope to create, a lifestyle that is both comfortable and enjoyable.

1.
1. Y.v Corporation. Japan
2. Y.v Skin / Hair Care Series
3. Hair Shampoo & Hair Conditioner
4. Plastic

2.
1. B&C Laboratories Inc. Japan
2. PHARSALA / Hair Care Series
3. Hair Shampoo & Hair Lotion
4. Plastic & Recycled paper

1.

2.

1.
   1. ATTENIR CORPORATION. Japan
   2. NATURAL GOMMAGE MASSAGE
   3. Cosmetics
   4. Plastic & Carton

2.
   1. ands Corporation. Japan
   2. PROGRAM CREAM
   3. Cosmetics
   4. Plastic & Glass & Carton

3.
   1. ands Corporation. Japan
   2. ATORREGE AD+ SERIES
   3. Cosmetics
   4. Plastic & Glass & Carton

4.
   1. AOIMAQUILLGE. Japan
   2. CATLLA / skin care series
   3. Cosmetics
   4. Plastic & Glass

5.
   1. CUEZ. Japan
   2. CQ / skin care series
   3. Cosmetics
   4. Plastic & Glass & Carton

6.
   1. House of Rose Co.,Ltd. Japan
   2. Wrinklaway Essence
   3. Cosmetics
   4. Plastic & Recycled paper

4.

5.

6.

1.
   1. SEPTEM PRODUCTS CO.,LTD. Japan
   2. Distris Hair Care Series
   3. Hair Shampoo & Hair Conditioner
   4. Plastic

2.
   1. MAX CO.,LTD. Japan
   2. HERBAL VALLEY GIFT SET
   3. Body Soap & Bath Essence
   4. Plastic & Carton

1.

2.

3. 1. MAX CO., LTD. Japan
   2. marie claire gift set
   3. Bath Essence & Soap
   4. Plastic & Recycled paper

4. 1. MAX CO., LTD. Japan
   2. FREE CHOICE GIFT
   3. Shampoo & Conditioner & Bath Salt & Soap
   4. Plastic

3.

4.

DESIGNAFAIRS
GERMANY

1 Managing Directors of designafairs
Nico Michler, Gerd Helmreich, Michael Lanz, Claude Toussaint

designafairs is a German-based, full-service design agency with studios in Munich and Erlangen. Together with a global network of material, engineering and research partners the team of 38 designers offers a wide range of international design services for well-known brands like Wella, LG, Siemens, Rubbermaid, Haier, WMF and many more.

In addition to packaging design, the company offers brand & design strategy, industrial design, and color & material design.

designafairs combines it's knowledge of clients' brands, the various markets, consumer requirements and diverse industrial sectors with design competencies; an interdisciplinary approach that yields sustainable, innovative design solutions. Depending on project type, the design team is formed individually and can include, for example, strategists, designers, psychologists, anthropologists and engineers.

designafairs is one of the largest design agencies in Europe. In European design awards ranking, the agency consistently holds one of the top positions - in fact, designafairs has been number one three times in a row.

The examples on these pages show the packaging design for the Siemens Gigaset Alessi Dect phone and the Wella High Hair styling series.

The Wella High Hair packaging design, in particular, is a good example of how knowledge of future consumer needs can lead to successful and sustainable design solutions. During the design process, various pieces of information concerning socio-cultural and fashion trends, as well as user insights were used to come up with innovative ideas for the new concept. This resulted in a timeless design with black TPE-parts forming a prominent key-visual for the series, which simultaneously enhances the ergonomics of the product.

The Wella High Hair packaging design has won six different international design awards: the World Star Packaging award, the Materialica Award, Cosmopolitan Prix de Beauté, the Good Design Award, iF product design award, and the Design Award of the Federal Republic of Germany in silver.

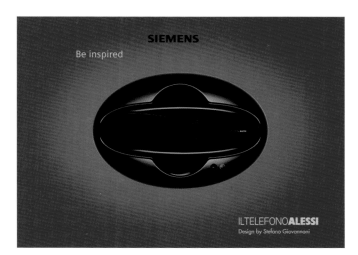

2 Name of Client / Siemens, Germany
  Brand Name / Gigaset Alessi
  Type of Product / Dect Phone
  Material / Paper

3 Name of Client / Siemens, Germany
  Brand Name / Gigaset Alessi
  Type of Product / Dect Phone
  Material / Paper

4 Name of Client / P & G Switzerland
Brand Name / Wella High Hair
Type of Product / Hair Styling Series
Material / Al, PE, TPE, Glass

DEPOT WPF

RUSSIA

Depot WPF Brand and Identity is local, privately owned branding agency. It was established in 1998.

Services
Core competences are brand strategy, packaging design and brand identity. Main services are brand strategy development, naming and verbal identity, packaging design, brand and corporate identity, brand guidelines creation.

Clients
Depot WPF has been working with local and international clients and brands such as Danone, TOTAL, Nestle Foods, Unilever, Gallina Blanca, Myllyn Paras Oy, Unimilk, Beeline, Aquavision and others.
It is the representative of Pan European Brand Design Association (PDA) in Russia.

Awards
Depot WPF is the winner of different awards such as Epica 2006, Golden Drum 2006, Kiev International Advertising Festival 2000, 2006, Moscow International Advertising Festival 1998-2006, EFFIE/BEST BRAND 2002.

Contacts
Russia 109004, Moscow,
Pestovsky pereulok, 16, bld. 2
Bussines-Center AKMA
Tel./Fax +7 (495) 363-2288
Email: info@depotwpf.ru
Web: www.depotwpf.ru

1 Dobra Kraska, Russia
Zvezda
Paints
Can
Depot WPF Brand & Identity

2 Uniservis, Russia
Chastnaya Gallereya
Confectionery
Cardboard, flow
Depot WPF Brand & Identity

3 Nestle Food, Russia
Nestle Classic
Chocolates
Cardboard
Depot WPF Brand & Identity

1. Dobra Kraska / Paints

2. Chastnaya Gallereya / Confectionery

3. Nestle Classic / Chocolates

4. Comilfo / Confectionery

5. Comilfo Lux / Confectionery

6. Ruzanna / Confectionery

**7. Uniway,
Russia
Gusto Vino**
Wine
Glass
Depot WPF
Brand & Identity

7. Gusto Vino / Wine

8. Pietro Coricelli Spa, Italy
Leonero
Olive oil
Glass, paper label
Depot WPF Brand & Identity

9. Technologies Francaises Alimentaires, France
Kenika
Tea
Cardboard
Depot WPF Brand & Identity

10. Nestle Waters, Russia
Saint Springs
Drinking water
Flow
Depot WPF Brand & Identity

11. Aquavision, Russia
BotaniQ Original
Smoothies
Flow
Depot WPF Brand & Identity

8. Leonero / Olive oil

9. Kenika / Tea

10. Saint Springs / Drinking water

11. BotaniQ Original / Smoothies

12. Trest B. S.A, Russia
Trest B
Georgian sauces
Glass, paper label
Depot WPF
Brand & Identity

12. Trest B / Georgian sauces

13. Maitre de The / Tea

14. Belaya Beryozka / Vodka

15. Liasson / Drinking yoghurt

16. Gartenz / Canned fruit and vegetables

13. Technologies Francaises Alimentaires, France
Maitre de The
Tea
Cardboard
Depot WPF Brand & Identity

14. Regata, Russia
Belaya Beryozka
Vodka
Glass
Depot WPF Brand & Identity

15. Unimilk, Russia
Liasson
Drinking youghurt
Flow sleeve
Depot WPF Brand & Identity

16. Paradox A.G., Austria
Gartenz
Canned fruit and vegetables
Paper
Depot WPF Brand & Identity

17. Multon, Russia
Live Rich
Juice
Cardboard
Depot WPF Brand & Identity

17. Live Rich / Juice

MINATO ISHIKAWA ASSOCIATES INC.

JAPAN

Minato Ishikawa Associates Inc.
minato@minatoishikawa.com
http://www.minatoishikawa.com

Welcome to our pages. My name is Minato Ishikawa. Formerly, I was an art director for a advertising agency (Young & Rubicam Inc.).
I did advertising for newspapers, magazines, and TV commercials. At present, I develop logos mainly for packages and graphic design for posters. As an art director, I worked with copy-writers as well as photographers and illustrators. Often, a lot of teamwork was required. Now, I do most of my own copies and illustrations
(getting help from others from time to time).In addition to my design work,
I have done many other things, such as set up stores and exhibitions, done stage art, been an instructor of design abroad, been a seminar instructor, production of music video, arts photography,object production, product development, introducing interview pages in magazines...and the list goes on. Most of my work has been related in some way to design. As well, I don't think I am exaggerating when I say my life itself is "design".
I believe design should be practical yet at the same time, appeal to our emotions and feelings. A good design is one that is born from the heart.I hope that our pages has helped you to learn more about my life, my work, and those around me.
We truly enjoy a LIFE OF DESIGN.

Minato Ishikawa

Kayoko Akiyama

Mio Ishimatsu

1.
1. Hakubaku Co., Ltd.
2. Wholly Organic japanese Noodle
3. Japanese Noodle
4. Japanese Paper

1. Krin Brewery Co., Ltd.
2. Maroyaka-kobo
3. Beer
4. Glass, Steel & Japanese Paper
5. Premium Beer

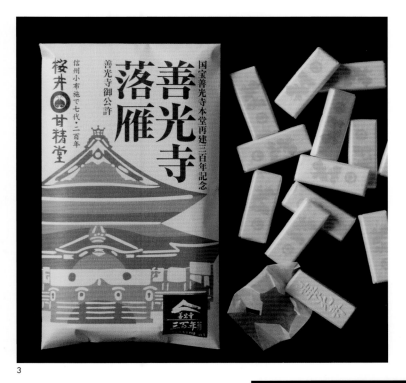

3 1. Sakurai Kanseido Co., Ltd.
   2. Zennkouji Rakugan
   3. Japanese-style Confection
   4. Japanese Paper
   5. Public Souvenir of Famous Temple

4 1. Benihana Foods Co., Ltd.
   2. Kuroegoma oil
   3. Cooking Oil
   4. Glass, Plastic & Paper

5 1. Edoya Records Co., Ltd.
   2. Matches of Edoya
   3. For sales promotion of CD
   4. Paper & Wood

6 1. Benihana Foods Co., Ltd.
   2. Kennbigenryou Cha
   3. Chinese Tea for diet
   4. Paper

7 1. Benihana Foods Co., Ltd.
   2. Grape Oil
   3. Cooking Oil
   4. Glass, Plastic & Paper

8 1. Sisen Syokuhin Co., Ltd.
   2. Tetujin Chin Kenichi's Shisennsai-no-umami
   3. Chinese Sauce
   4. Glass, paper & Steel
   5. Original Source of Famous Chinese Chef

9 1. Benihana Foods Co., Ltd.
   2. Grape Oil
   3. Cooking Oil
   4. Glass, Plastic & Paper

B&G
FRANCE

From the left to the right are : Yannick Soubrier, Creative Director, Cécilia Tassin, Head of Strategic Planning, Elodie Orieux, Client Services Director, Nicolas Chomette, General Manager, Nicolas Julhiet, Client Services Director and Mélanie Gransart, Creative Director.

Like any living organism, a brand needs to evolve and adapt to constantly changing environments. With 45 specialists of all design disciplines, B&G is an international design agency based in Paris. Our integrated range of marketing, research and design services helps you to develop your identity, your packaging, to anticipate changes and inspire your innovation process.

Guerlain/Packaging Design

Danone Waters/Brand Identity and Packaging Design

Carrefour Group/Brand Identity and Packaging Design

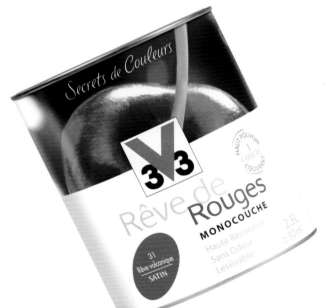

V33/Packaging Design

Danone Waters/Brand Identity and Packaging Design

Foods International/Packaging Design

MILK

GREECE

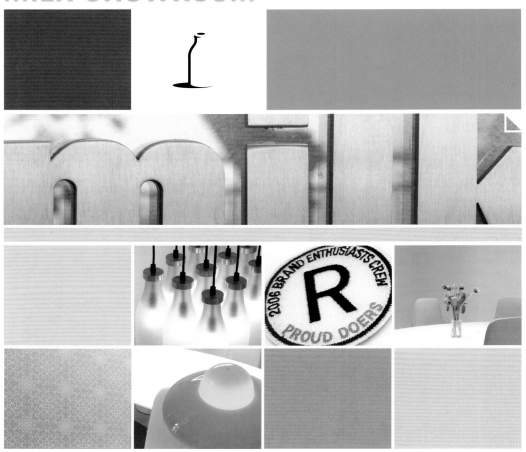

1. 1. PAPASTRATOS
   2. ANNIVERSARY EDITION
   3. Cigarette Packaging

2. 1. TIM HELLAS
   2. F2G
   3. Connection Pack
      (Telecommunications)

3. 1. VODAFONE HELLAS
   2. CU
   3. Connection Pack
      (Telecommunications)

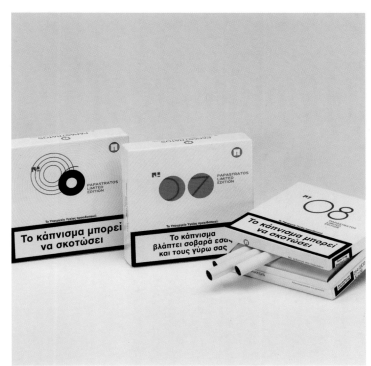

1.

2.

3.

4.

6.

5.

4. 1. SARANTIS S.A.
    2. ORZENE BEER SHAMPOO
    3. Hair Care Products

5. 1. SARANTIS S.A.
    2. TRILET
    3. Liquid Dishwashing

6. 1. SARANTIS S.A.
    2. B.U.
    3. Eau De Toilette
       Fragrance Series

7. 1. ATHENIAN BREWERY S.A.
   2. AMSTEL BEER
   3. UEFA Champions League Special Edition Packs

8. 1. FINE GREEK FOODS
   2. EF
   3. UExtra Virgin Olive Oil

9. 1. MEDITERRA WINERY
   2. CURRICULUM VITAE
   3. Dry White Wine 750 ml

10. 1. MEDITERRA WINERY
    2. VINO NANO
    3. Semi Dry Red Wine 187 ml
       Semi Dry White Wine 187 ml

7.

8. 9. 10.

VILLEGERSUMMERSDESIGN

U.K.

Our Anglo-French partnership has over seven years experience designing for clients in Europe, for industries including confectionary, luxury perfumes and kitchenware.

Over the years, we have gained the trust of several major brands, increasingly keen to collaborate with pragmatic, and level-headed designers.
These include Yves St Laurent, Issey Miyake, Typhoon, Van Cleef & Arpels, Burnt Sugar, Boucheron, The Body Shop and Givenchy.

Our ability to understand a companyís needs, identity, and objectives has resulted in successful designs, and sustained relationships with many of our clients.

We believe it is important to be flexible, approachable and reactive. And as well as being reliable, we want to make sure we are easy to work with.

Our services include consumer product design, packaging design (structural and graphic), gift with purchase and box set design as well as corporate graphic design.

Starting from a sketchbook of ideas, we will work closely with your product development team to generate concepts, providing elevations and three dimensional renderings, through to dimensioned drawings for prototyping.

Your feedback is welcome at every stage of the process, ensuring a result which will match your requirements and exceed your expectations.

Here is what a few of our clients have to say about us:

"If your project deadline is far enough, if your budget is large enough, if you are not really looking for fresh and out-of-the-box ideas, then you could probably pass on this agency. My reality was more brutal and working with villegersummersdesign always guaranteed soft landing projects. Internationally minded, always interacting with a smile. Simply talented."

NV, product manager, Issey Miyake Perfumes

"We have worked with Vincent and Angela for nearly three years now and have been amazed by their creativity and professionalism. They have listened carefully to our briefs and interpreted them with real flair.
By working with villegersummersdesign, our brand look now has much more impact which has had a direct effect on gaining more product listings with our customers (Harvey Nichols, Waitrose, etc)
Aside from the excellent work they do (including always managing to meet even the tightest of deadlines), Vincent and Angela are a real pleasure to work with."

JC, Director, Burnt Sugar

"villegersummersdesign have played an integral part in defining the ëGiftedí brand label at Typhoon over the last 3 years. From the early beginnings of a modest range, they have designed the strong graphic packaging looks which have been fundamental in turning the brand label into the £1,000,000 turnover business it is today.
Our ëGiftedí clients in the UK and overseas have selected ëGiftedí to range in their shops because of the great packaging which ëshoutsí from the shelvesí in what is a very competitive market.
My experience of villegersummersdesign has always been that of a very organised and professional design agency, but also flexible and easy to work with: attitudes that are often missing from so many London agencies. They always provide a prompt and reliable service, often working to very tight deadlines."

HB, NPD manager, Typhoon

Van Cleef & Arpels Zanzibar Fathersí Day Gift Box

1.Yves Saint Laurent Opium 30ml Purse Spray 2.Burnt Sugar UK confectionary company 3.Van Cleef & Arpels Valentineís Gift with Purchase 4.Yves Saint Laurent Cinema limited edition purse spray

GRAHAM HANSON DESIGN
U.S.A.

Studio Exterior

Library

Graham Hanson Design is an internationally-recognized design agency active in all areas of strategic visual communications, including corporate identity, branding, print communications, packaging and advertising; Web design and development, electronic presentations and other interactive media; exhibit and event design, architectural graphics and signage, and wayfinding.

The agency works with a wide range of clients, from corporations, retail establishments and real estate developers to museums, architectural firms and cultural organizations. The diversity of our client base and our broad range creative service capabilities reflect our belief that effective and compelling solutions are driven by collaborative client relationships and a clear understanding of context.

Graham Hanson Design consists of a diverse group of designers, writers, project managers, content developers and producers and is located in New York City's Flatiron District, just off Madison Square Park.

For more information, please visit www.grahamhanson.com.

Skin Therapie
The Skin Therapie line of cosmeceuticals, from the labs of International Cosmetic Surgery, launched internationally in select department stores in September 2006. Graham Hanson Design developed strategic brand development, packaging, marketing collateral and point-of-purchase.

International Cosmetic Surgery / Design: Graham Hanson, Dorothy Lin

STUDIO 360
SLOVENIA EUROPE

Vladan Srdic, CEO & Creative director

CONTACT US!

Address: Kotnikova 34, SI-1000 Ljubljana, Slovenia, EU

Phone: +386 (0)31 847 261

E-mail: office@studio.360.si, studio@thesign.org.uk

Web: www.studio360.si, www.thesign.org.uk

STUDIO 360 is a company providing integrated solutions in the fields of architecture and branding. The combination of two- and three-dimensional expertise contributes to strategic brand development, efficient results and client satisfaction.

WE THINK 360: branding department offers comprehensive solutions for advertising, illustration, packaging, graphic and web design. Our philosophy is communication in a fresh and sophisticated style: simple, clever and always with a twist. We are focused on efficient solutions and dedicated to obtaining a higher value for your brand - leaving nothing to chance. Every project is the most important to us: combining the principles of aesthetics, idea and function, we are commited to creating strong concepts and added value for our clients.

WE DO 360: architectural department focuses on flexible design concepts and innovative building techniques that improve the quality of living. Every building receives a comprehensive solution: analysis of context, client's brief, investment and construction technologies. We aim to achieve more with less, and creatively transform constraint into opportunity. Studio 360 provides services encompassing, but not limited to, urban planning, exhibition set-up, building exteriors and interiors. Design is based on simplicity, attention to detail and intelligent use of materials.

STUDIO 360 projects have been featured in distinguished books and magazines and have won many awards at domestic and international competitions. Ten years of experience and dedication ensures excellence. We are flexible, fast and capable of producing total solutions both graphically and architecturally.

SELECTED CLIENTS: Ericsson, Akton, Parsek, Difar, Kalliste, Krka, Zavarovalnica Maribor, Lassana, Mediamix, BSL, CCEA, Val Navtika, Magdalena Festival, Mladina, MGLC Gallery, Fabula, Ministry of Culture of RS, Fabriano, Rambo Amadeus, Katarina Venturini, Gorisek-Lazar, Theatre Dusko Radovic, Torpedo theatre company, Vrooom, Life Dust, Wiener Staedtische...

1. Client name: Difar d.o.o.
2. Country: Slovenia
3. Year: 2006
4. Brand name: Prodiet
5. Type of product: Package for diet food

STUDIO360

1. Client name: RTV Slovenija
2. Country: Slovenia
3. Year: 2006
4. Brand name: Gorisek-Lazar
5. Type of product: CD cover

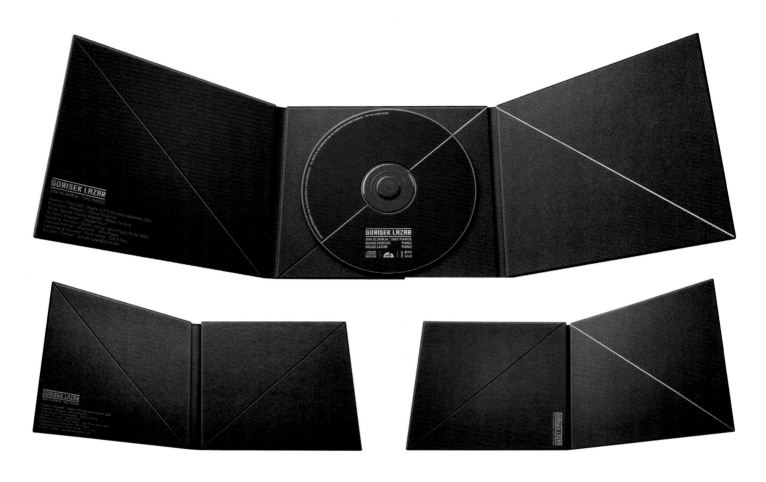

STUDIO360

1. Client name: B92
2. Country: Serbia
3. Year: 2005
4. Brand name: Rambo Amadeus
5. Type of product: CD cover

TRIPLE 888 STUDIOS
AUSTRALIA

Established 20 years ago, Triple 888 Studios has grown to provide a very broad range of creative design and strategic marketing services.

Our seven strong design team have produced many award winning creative solutions to clients needs – from packaging (import and export), brochures, corporate images, displays, advertisements and web sites.

Our services also include Marketing Strategies, Media Planning, Public Relations, Direct Marketing, Event Management, Television and Radio Production, Sales Promotions and Database management for specialised clients.

Triple 888 Studios provide creative artwork services of the highest standards with meticulous attention to detail, showcasing each client's products in the best way possible.

As a successful business, our services are employed by local and international clientele and has prepared language specific artwork for Europe, United Kingdom, Asia Pacific, South Africa and Gulf regions.

Local clients consist of manufacturers operating in a wide range of industries including pharmaceutical, industrial, housing, cosmetics, automotive, homewares and consumables. These clients include GlaxoSmithKline, BP-Castrol, DEB Australia, Felton Homes, Royal Australian Navy, Sheldon and Hammond, Jonsa Ellies, Aspen, Miller's Retail, Southern Cross, Le Mac Enterprises, Form-Tek, Universal Publications, PT Hydraulics Australia, Form-Rite & Markets Unlocked.

We maintain a strong consultative business partnership with our clients, with the objective of always exceeding their expectations.

This flair for innovation has lead to several other awards recognising Triple 888 Studios excellence. These include:

- National Print Awards
- Australian Catalogue Awards
- Australian Packaging Awards
- Summit Creative Awards
- Supplier Recognition Awards
- Premier Print Awards (worldwide)
- Creative 33,34 & 36 Award
- Western Sydney Arts Business Awards

We constantly strive to achieve the highest standards of services and in doing so, have created long lasting partnerships with our clients.

Michael Quan
Managing Director

1 Promax/ Plastic can

2 Sirtex Sir-Spheres Pty Ltd
Accessories pack

3 Audio Axis
Car MP3 Player

4 Promax
Plastic can

5 Sheldon and Hammond
  Ikasu Knife Block

6 Sheldon and Hammond
  Kamati Shinjo Knife Block

7 DFF Beverages/ Ice Black Espresso Coffee

8 Sheldon and Hammond/ Silice Knife Block

9 H2O4K-Water for Kids Water Bottle

10 Europe Asia Trading Kookaburra Ridge Wine

11 Healthy Vegie Bite/ Vegie Spread

PATRICIA CATALDI BRANDING & DESIGN

BRAZIL

A company focusing on results, Patrícia Cataldi operates with a nucleus specialized in strategic design. The company's scope comprises projects in the areas of branding, corporative identity, visual communication, packaging and graphic projects as well. The Patrícia Cataldi design nucleus work concept is based on the modernest identity project management practices. The nucleus possesses a network of renowned partners working in the most varied areas that support the feasibility of the construction of successful brands. Our differential is the differentiated treatment focused on the generation of long term results and uniqueness for every one of our clients.

patricia@patriciacataldi.com.br

Praça Vilaboim 78 6º andar Higienópolis
01241 010 São Paulo SP Brasil
Tel: 5511 36686572 Fax: 5511 36673462

IDP International Design Partnership
www.idpweb.com

1. ALMAPBBDO/AMBEV - Brazil
2. Antarctica
3. Pilsen Beer
4. Label and aluminium can
5. Design by Roger Testa

KEIZ OBJ WORKS INC.
JAPAN

My name is Kojiro Tomoeda

KEIZ OBJ WORKS INC.
Kojiro Tomoeda
tomoedakojiro@mac.com
tomoeda@mrc.biglobe.ne.jp

Home Page
http://homepage.mac.com/tomoedakojiro/

Blog
http://blog.goo.ne.jp/tomodesmo

Heart Village shop Home Page
http://homepage.mac.com/tomoedakojiro/heartvillage/Menu19.html

We keep the design that a heart is comforted in mind.
I want to stick to the design because it is the age when things are left over.
Happiness, a shape good point, prettiness are necessary for which age as well.
Because I want to make what is close to a person's feeling, the theme of our design is "a feeling good point".
Moreover, it grapples eagerly, and the design that a Japanese traditional element was adopted wants to go, too.

Digital Picture. Illustration by ©Kojiro Tomoeda

Smithsonian's America: An Exhibition on American History and Culture ©1994 Smithsonian Institution

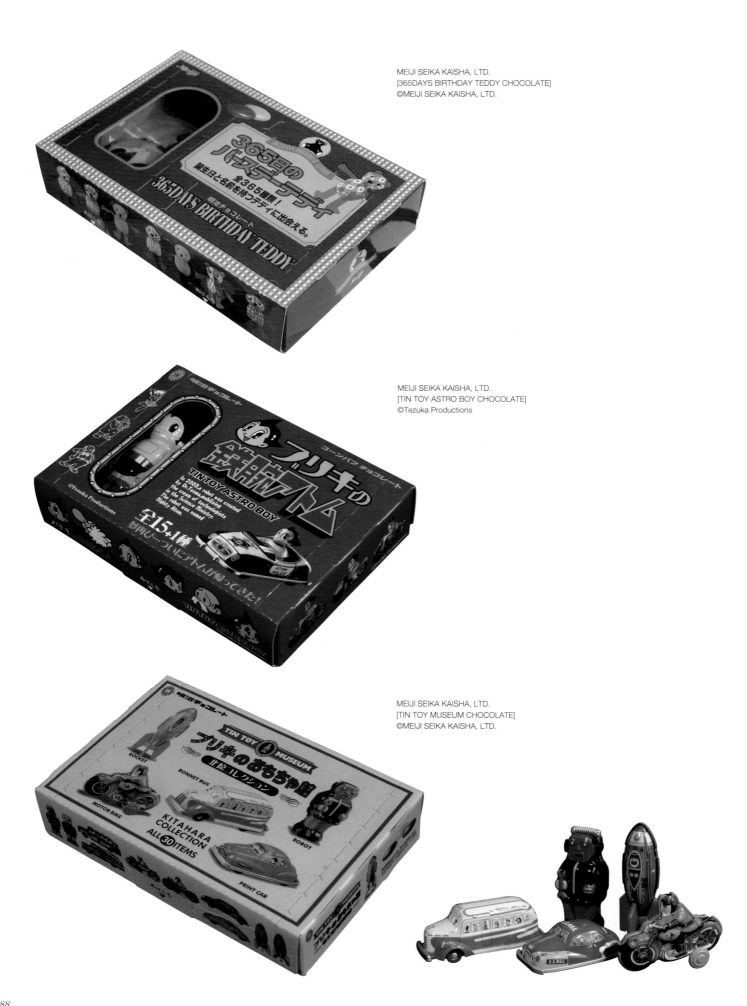

MEIJI SEIKA KAISHA, LTD.
[365DAYS BIRTHDAY TEDDY CHOCOLATE]
©MEIJI SEIKA KAISHA, LTD.

MEIJI SEIKA KAISHA, LTD.
[TIN TOY ASTRO BOY CHOCOLATE]
©Tezuka Productions

MEIJI SEIKA KAISHA, LTD.
[TIN TOY MUSEUM CHOCOLATE]
©MEIJI SEIKA KAISHA, LTD.

Amway's distributor of the world maximum Mr. Kaoru Nakajima's Private Brand [Heckel] Desk Clock & Alarm Clock   ©Heckel

Maboroshi Stationery [Ankodama Sweets]　©Toy Museum

Hananoka Brewing Co. [Mosaku Shochu Special Label] ©Hananoka

CRAVE
U.S.A.

crave™

Crave is a brand identity & packaging design firm for food & beverage companies. We help create, refresh and package food & drink brands.

To win over today's savvy shoppers, nothing short of mouthwatering visual temptation will do. At Crave, we specialize in creating food & beverage packaging that entices its target audience. We craft brand personalities that instantly communicate with clarity and style. Our sole focus is to build brand solutions that attract and connect with consumers—visually delicious brands people crave in the blink of an eye.

1
1. Organic Cottage LLC, USA
2. Organic Cottage
3. Organic foods from India
4. Clear plastic bag

2
1. The Sugar Plum Fairy Baking Company, USA
2. Sugar Plum Fairy
3. Premium, all-natural frozen desserts
4. Coated paperboard carton

3
1. Southern Specialties, Inc., USA
2. Southern Selects
3. Premium, fresh produce
4. Clear plastic bag

4
1. IQ Beverage Group, USA
2. IQ H2O
3. Vitamin enhanced water beverage
4. Clear plastic bottle with coated label

4

5
1. Atlantis Foods, Inc., USA
2. Atlantis Premium
3. Fresh, all-natural soup
4. Frosted plastic container with coated label

UNIQA C.E.
RUSSIA

UNIQA C.E. is an independent consulting and brand design agency that has worked on development of major Russian brands for the past 6 years.

Why UNIQA or what is so unique about our agency?

We placed a strategic brand consulting agency and a professional studio of brand design under one roof. This combination enables our professionals to see the strategic perspective and suggest the best solution. We're not trying to play the existing market, but rather boldly challenge the future. This allows us to create quality brands effectively from the ground zero.

Today UNIQA C.E. agency is the leader in alcohol brand creation in Russia and has significant experience working with leaders of the industry both in our country and abroad. We are also updating our knowledge of glass and polygraph industries. Our team is constantly experimenting and delivering new innovative solutions for our clients.

At the same time we are not overly specialized. Our portfolio includes several large projects dedicated to real estate development, FMCG, and other areas of business.

UNIQA C.E. became a full member of the European Association of Brand Design PDA in 2006.

We love our work. We believe in its importance. We create immaterial and deliver real profits for our customers. We develop strong and solidly positioned brands. We provide the information and insight our clients need to operate from a position of knowledge. That knowledge creates competitive advantages – advantages that translate into UNIQA stories of success.

1.
Russian Alcohol, Russia
Vodka Marussya
Glass

2-9
Russian Alcohol, Russia
Project Green Mark
Glass

10

11

10-11
HUSKY, Russia
Vodka HUSKY
Glass

12
Ferein, Russia
Vodka Shustov
Glass

13
OST Group, Russia
Vodka Moskovskaya Guberniya
Glass

12

13

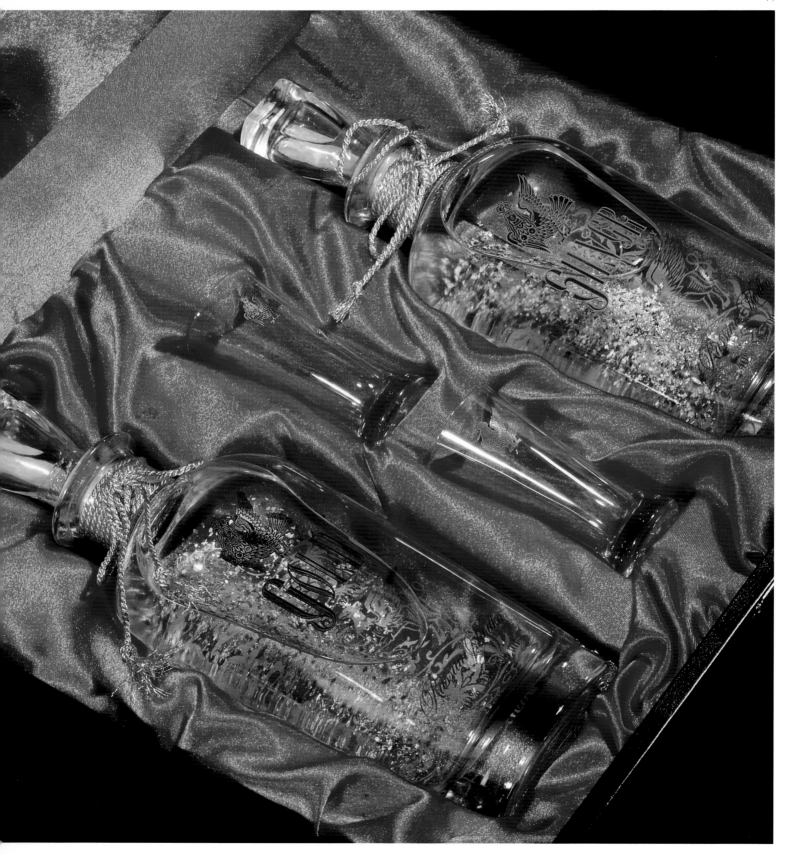

14
Glavvino, Russia
Vodka Gold and Silver
Glass

15
Russian Alcohol - Ushba, Georgia
Djera
Glass

16
Russian Alcohol, Russia
Vodka Yamskaya
Glass

17
Russian Alcohol - Ushba, Georgia
Vodka Ushba
Glass

18
Darsil, Russia
Souses GEO Collection
Glass

TANGRAM STRATEGIC DESIGN
ITALY

1981
Tangram Strategic Design was founded. As the name implies, we work in strategic design, and as the facts show, we do it successfully.

Our Philosophy
One client – one project.
One problem – one solution.
This is how we work. Every situation deserves an individual approach, because in the complex world of marketing, no two situations are alike.
A creative idea does not come from a moment's intuition, but from market studies, detailed analyses, and the formulation of a strategy. All with the constant collaboration of the client.

Our mission
Many agencies impose their own particular "style"; our aim is to make design serve companies to achieve a concrete objective.
Creativity, innovation and strategy. These are the ingredients of quality marketing, in a market where quality is synonymous with success.

Our services
Corporate & Brand Identity, Naming
Packaging Design
Corporate Literature & Editorial
Information Design
Web Design (for extending the values of the brand online)
Exhibit & Retail (to give the brand a physical nature).
Our prerogative is to have experts specialised in all sectors, in order to treat every area as a dedicated study.

Our network
Tangram Strategic Design has built up a network of collaborators that include companies and professionals specialised in:
- Industrial Design
- Architecture
- Psycholinguistic and Motivational Research
- Strategic consulting
- Public Relations.

These collaborators form the base of our network. For every project, a specific working team is created, and co-ordinated by one of our strategic heads to ensure that the values of the brand remain the centre of attention throughout the whole process.

Our methods
Good methodology is at the heart of a good project, just as problem-solving is at the heart of our creativity.
For brands in difficulty, we have created Revive, a new method that, with the support of psycholinguistics, looks at the past and present of the brand, in order to create a new future for it.

Costs/benefits
There are those that pride themselves on their price, there are those that consider price synonymous with quality; and then there is Tangram Strategic Design.
Thanks to its external network of collaborators, the structure does not create additional costs for the client, and every project is assessed according to the resources actually used. Large companies know how best to use our services, and they appreciate the results.

1

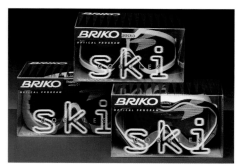

2

1, 2, 3
Brico, Italy
Briko, underwear / ski goggles / helmets bike
Cardboard box / PET / cardboard box
Packaging system restyling

4

5

6

7

8

4
Chiquita Italia, Italy
Chiquita Tropicals, fruit juices
Tetrapak
New packaging system

5
Castello di Udine, Fabbrica friulana di birra, Italy
Castello, bier
Glass, labels paper
Packaging system restyling

6
Barilla, Italy
Mulino Bianco, bread
Cellophane wrap
New packaging system

7
Davide Campari Group, Italy
Biancosarti, aperitif
Bottle, labels paper
Packaging system restyling

8
Danone Group, Italy
Saiwa Dolcezze del Mondo, biscuits
Cardboard box
New packaging system

9

10

11

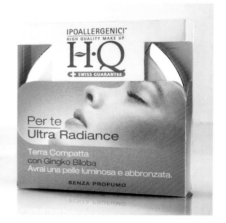

12

9
Interdis, Italy
Dimeglio, toiletries
PET bottle
New packaging system

10
Damiani Italia, Italy
Bliss, jewels box
ABS box + nester varnish
New packaging system

11, 12
Paglieri Sell System
HQ, mini-doses / make-up
Plastic / cardboard box
New packaging system

13
Alfawassermann
Triene, burn cream
Cardboard box
New packaging system

13

im2design
POLAND

If there were as many recycling sites as there are design agencies claiming they are different, "melting glacier" would remain an oxymoron. There'd probably be more truth in saying that we're all the same by trying to convince the client that we're all different.

1.

1.

im2design / POLAND
Money Poetry
self- promotional piece
paper & metal
pages designed of money

2.

trick
Erotic Tea Shop
tea
paper
tea for men and women

3.

Trick
COENRAAD Chocolates
Chocolates
Paper
Luxury chocolates

4.

Divon
Aqua Diva
Mineral water
Glas, paper, cork, string
Special edition of mineral water

ONC DESIGN STUDIO

JAPAN

Address 3-2-17-214 Nakazaki-Nishi Kita-ku, Osaka 530-0015
Phone/Fax 06-6375-1301
E-mail onc_nishino@extra.ocn.ne.jp
Website http://www9.ocn.ne.jp/~onc/

Osamu Nishino

Minako Mihara

ONC Design was established in 1982, since then we have been working with well-known confectionary companies to create many long seller items.
It is our motto to understand client intentions and the distinctive characteristics of products at first to then create designs that convey a meticulous care for details.
We are known for the effective use of calligraphy, decorative images with a variety of patterns and image embossing.
We aim at creating timeless package designs.

1
1. Ginso Corporation
2. Noble Neige
3. Ice Pudding

1. Morozoff Limited
2. Peerage
3. Chocolate

3 1.Morozoff Limited
   2.Gelato alla Frutta
   3.Sherbet

4 1.Morozoff Limited
   2.Premium Chocolate Selection
   3.Chocolate

5 1.Ginso Corporation
   2.ASUKA
   3.Castella

6 1.Toraku Foods Co.,Ltd
   2.Kyo no Omaccha Chocolat
   3.Maccha Chocolate Cake

7 1.Ginso Corporation
   2.Aska
   3.Castella

3

4

5

6

7

8   1. Ginso Corporation
    2. Papillon D'or
    3. Sweets

9   1. Ginso Corporation
    2. Mango Jelly, Pudding
    3. Sweets

10 1. Ginso Corporation
    2. Royal Jersey Pudding
    3. Pudding

11

11
1. Marukan Vinegar Co.,Ltd
2. Asa no Kajitusu
3. Vinegar

12
1. Naris Cosmetics Co.,Ltd
2. Cholestefree, Bidens Pilosa
3. Health Foods

13
1. Ginso Corporation
2. Mizuyokan, Wafu Jelly, Fleur Brillant
3. Summer Sweets Gift

12

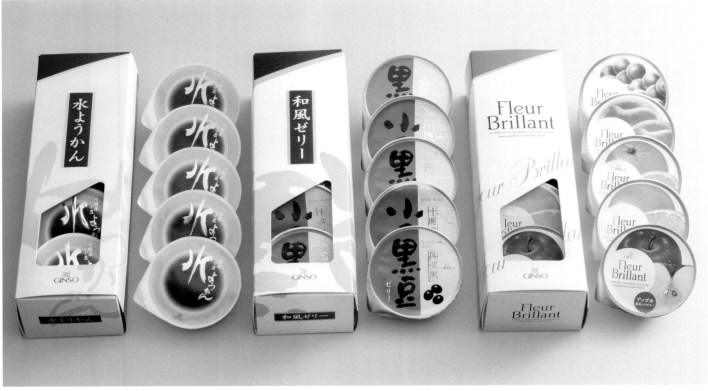

13

FiF DESIGN

THAILAND

FiF DESIGN is a Branding and Identity Design Consultancy based in Bangkok, Thailand. We help our clients build brands and businesses by delivering new experience in strategies, and design executions. Our multidisciplinary team offers compelling creative and extraordinary solutions ranging from branding issues, design research, product design, retail environment design, packaging design, and communication design.

Great idea comes from great people. With our talented, passionate, and energetic people, we are conscious what we are concentrate on. For the mixer of multidisciplinary team with both sides of equation, arts & science, we all work together, discuss, brainstorm, explore, and experiment many possible way to bring new experience to our client brand, product, service, or space.

Design is the great story to make different to business and people life. We believe that people generate experience from the story they perceive and from what they can sense; see, touch, smell, and taste. Since people look for experience to fulfill their dream and match their lifestyle, it is the heart and soul of our thinking to create the meaningful story and express its identity consistently to the right contact points. This is what we believe and focus to create the extraordinary design solution for the success of our client's business and, of course, for better life.

S&P Syndicate Public Company Limited
Moon Cake Packaging
Metal Sheet (Press Work)
Chinese papercut representing chinese ritual (luck) / FiF DESIGN

Unilever Thai Trading Limited
Knorr CupJok
Cup with shrink wrap label / Pouch
Devilish fun experience / FiF DESIGN

Parker & Morgan
Spa Packaging
Bottle
Serenity / FiF DESIGN

Sweet Treat (Opposite Page)
Chocolate Packaging
Wooden Box
Special gift with love / FiF DESIGN

Nestle (Thai) Ltd
Sterilized Milk Packaging
Packaging Label
Hygenic Premium / FiF DESIGN

White Tiger
Liquor Packaging
Packaging Label
Bold / FiF DESIGN

Unilever Thai Trading Limited
Knorr CupJok Hong Kong
Cup with shrink wrap label / Pouch
Emperor Taste / FiF DESIGN

PTT PCL. (Opposite Page)
Lubricant Packaging
PE Blow Molding Bottle
Dynamic Power / FiF DESIGN

STORM CORPORATE DESIGN LTD.

NEW ZEALAND

Rehan Saiyed, a multidisciplinary designer committed to solve design problems incorporating a number of disciplines form print, packaging, new media and architecture.

I believe that to be successful, the communication of a company or product must be conveyed through appropriate and effective design based on marketing strategies. The development of global image created by a coherent design assures the success of any product.

For more than ten years, I have created this global image through graphic, product, new media and architecture design. I approach the design with the intent to personalise a company or product's image within its market.

Based in New Zealand, Storm Corporate Design is dedicated to serve the design needs of an international clientele - in print, packaging, new media and architecture.

1.   Cadbury, India
Cadbury Assorted Chocolates
Chocolates
Duplex Board
Festive Packaging / Storm Corporate Design

2.   Bravo, New Zealand
Bravo Deodorant
Toiletries
Designed product and brand name, silk screened on the bottle

1. Cadbury

2. Bravo Deodorant

3. Silver Spring Mineral Water

4. Kleen Out

3. S&G India Ltd., India
Silver Spring
Mineral Water
Silver metallic printing on Silver Foil

4. Manhattan Technologies, India
Kleen Out
Washing Gel
Process colour with Blue foil on Chromo paper

5. Punjas Ltd, Fiji Islands
Fiji Style
Food
Process colour on Chromo paper

5. Fiji Style Chutney

6. Dinners On Ice

6.   Dinners on Ice, New Zealand
       Dinners on Ice
       Food Packet
       Process colour on Chromo paper

7 & 8.   Sentimenz Ltd, New Zealand
       Sentimenz
       Gift Packs
       Process colour on Duplex board

7. Sentimenz Gift Packs

8. Sentimenz Gift Packs

VIE DESIGN STUDIO
INDONESIA

In the year of 2002 in Bandung City-Indonesia was a start of our business in the area of Creative Branding and Strategic Communication.

Initiated from our passion in creating design that's effective, communicative with high artistic value, we try to accomodate our client's need of maximum result. In the process of creating a design, we always try to do a personal approach because we believe that the key to success of a design is to understand and to learn our client's business.

Other than that, we also invite our client to participate and to enjoy in the process of creativity, aligning our minds through the design that we created.
By then, perception about a good design will be understood by our client and we expect that our client could give a good appreciation of the art of communication.

Vie design offers an integrated solution which includes: coorporate & retail identity, branding & packaging, annual report, company profile, digital imaging, environmental graphic design.

The Showroom

1. Client Name. Bakmi Pandanaran (Indonesia)　　　Material. Ivory Paper
   Brand Name. Bakmi Pandanaran　　　　　　　　Theme. Traditional Noodles of Semarang City
   Type Product. Delivery Box & Bag

2. Client Name. Ny. Liem Cake (Indonesia)
   Brand Name. Ny Liem Kursus Kue & Masakan
   Type of Product. Box & Bag
   Material. Artpaper
   Theme. One Month Baby Celebration
   (Chinesse Tradition)

3. Client Name. Tunggal Inti Kahuripan (Indonesia)
   Brand Name. Cheese Milk Biscuits
   Type of Product. Plastic Package
   Material. Plastic
   Theme. Delicious Biscuits for Gathering

4. Client Name. Tunggal Inti Kahuripan (Indonesia)
   Brand Name. Biscuits
   Type of Product. Plastic Package
   Material. Plastic
   Theme. Strong Like The Dragon

5. Client Name. Gizifood Prima (Indonesia)
   Brand Name. Pisa Pizza Restaurant
   Type of Product. Delivery Box
   Material. Corrugated Paper & Ivory Paper
   Theme. Enjoy Your Life with Pisa Pizza

6. Client Name. CV. Global Ice (Indonesia)
   Brand Name. Royal Gelato
   Type Product. Cup Label
   Material. Art Paper
   Theme. - Premium Class
           - Polar Bear

PEARLFISHER

U.K.

Pearlfisher is a leading independent design agency owned and managed by three partners. The company was founded in London in 1992, and we opened our second studio in 2003 in New York. We operate internationally and our biggest clients are from South Africa, Asia, America and Western Europe.

Pearlfisher creates future desire for brands.

Our philosophy is based on the concept of truth and desire: uncovering the essential truth at the heart of a brand and expressing it in a way that consumers will find desirable as we move into the future. To do this, we need a clear perspective on the future, so we proactively seek to identify the cultural shifts that will impact consumer behaviour, and we approach every project with an awareness of its potential future impact on the environment.

Our approach can be broken down into three stages:

Anticipating desire: seeking to understand future consumer behaviour through our transatlantic insight program LifeModes. Using our network of opinion formers – from inventors and experts to artists and streethawks – we discuss, debate and deliberate changing human desires.

Interpreting desire: understanding and illuminating what these changing desires mean in the context of a particular brand. Our planning team uncover the big ideas and strategies that will help brands navigate the future.

Creating desire: translating our ideas and insights into a desirable expression of the brand. Our design and realisation teams can create identity, packaging, literature, environment, film and digital solutions for our clients.

The result is something we call good design: desirable solutions for the long-term success of brands created with an awareness of the future needs of the planet.

Our client list includes some of the world's most iconic brands. We work well with established brands, challenger brands and, increasingly, brands that put ethics before profits:

Absolut, Coca-Cola, The Edrington Group, Fortnum & Mason, Green & Black's, Havana Club, InBev, Innocent Drinks, Liberty of London, Pizza Express, Soho House, Unilever, Waitrose, The White Company and Woolworths South Africa.

Pearlfisher is managed hands-on by the three original partners, Mike Branson (Managing Partner), Jonathan Ford (Creative Partner) and Karen Welman (Creative Partner). Our 50-strong team in London and 10-strong team in New York are structured around our three stage integrated approach and encompass the talents of planning and insight specialists, 3D, digital and graphic designers, materials and environmental experts, writers, film-makers, animators and artworkers. In addition, every client is supported by a full account management team.

pearlfisher

BRAND AID

UKRAINE & RUSSIA

BrandAid
brand innovation company

BrandAid is the unique company who creates brands. A unique, because we consolidate under the one roof both strategic and designer parts of brand development process. Usually those functions are separated. However we decided that the market needs the full service in the area of brand creation and development. That is why we conduct market research to determine the level of market development and provide consulting, helping companies to get the very right positioning. Besides, we totally "dress in" the brand: create the name, logo, package, CI and other elements of the brand, that help customers to choose us.

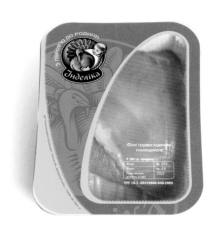

1. Ramburs. Ukraine
2. Indelica
3. Poultry
4. Plastic tray & polyetilen film with flexographic printing
5. Prosperity of my family is the source of my stability
It is essential for me to protect them and to take care of them
Turkey meat - is a valuable product that helps me to show my love and concern about my relatives

1. Bunge. USA
2. Oleyna
3. Vegetable oil
4. Plastic bottle. The label is prodused from a paper with offset printing
5. Dominate the category: both quantitatively and qualitatively
Dominate the trend (from "Ukrainian natural" to "modern-&-healthy")
Enrich the category with new benefit (emotional aspect that has not been communicated previously): "it's all about pleasure".

MIDNETE OIL
THAILAND

Mongkolsri Janjarasskul / Design Director

Midnite Oil is a multi-disciplinary design firm, founded in 1993 by an ex advertising agencies team who graduated from USA. It was a sister company of The Magic Wand, the first service bureau of Thailand offering high resolution input and output services including film recording and color separations. Working side by side with The Magic Wand, Midnite Oil team had benefits over her competitors as they had strong knowledge in prepress which was very important for graphic designers. With the tender age of two, she received 3 retouching awards from Gallery of Excellents, a high end computer graphics competition, held in Orlando, Florida, USA. Although the 1st Prize and 3rd Prize in Photo Retouching/Manipulation category were rewarding, the "Best of Show" prize which was voted by the general audience, from 1st Prize winners of 10 categories, was a truly gratifying accomplishment.

Started as a graphic design firm offered only design related to printed matter, nowadays the services cover all about Branding. Not only prints which includes graphic design, packaging design and advertising, Midnite Oil also offers shop and display, website and interactive design works. The account she handled including international brand clients such as JVC, McDonald's, Clairol's, Schweppes, EMI, Cannon, Prudentials, Amway and etc. At the moment Midnite Oil started to gain acknowledge from international markets. The work came from overseas such as Macau and The Netherlands.

Although having a real passion in design and creativity, Mongkolsri Janjarasskul, Design Director and a co-founder of Midnite Oil, believes that an exceptional design work must also push sales volume and accurately portray the client's branding, which is a primary function of commercial art.

* The name "Midnite Oil" comes from the phrase "burning the midnight oil", or to work late into the night.

1. ZhongHuaDaoBao

Client Name:	ZhongHuaDaoBao/Thailand
Brand Name:	ZhongHuaDaoBao
Type of Product:	Pearl extract
Material:	Art Card paper
Theme or Concept:	As the product's benefits are for health and beauty and imported from China, the packaging gift set needed to portray a look and feel of a luxurious and precious gift of extraordinary beauty.

1. ZhongHuaDaoBao/ Thailand, ZhongHuaDaoBao, Pearl Extract

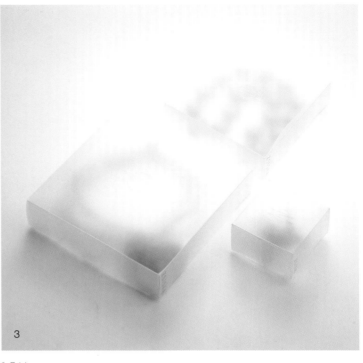

2. Zummer

Client Name:	Zummer/Thailand
Brand Name:	Zummer
Type of Product:	Tomato Juice
Material:	Tin
Theme or Concept:	Zummer product line originated with Chinese Tea utilizing a logo and motif of traditional Chinese brush work. When this assignment came to us, we opted for incorporating an oriental watercolor approach to portray a fresh look yet still maintain original essence of the brand image.

3. Zsiska

Client Name:	Siska Design/Thailand
Brand Name:	Zsiska
Type of Product:	Fashion Jewellery
Material:	Matt PP sheet
Theme or Concept:	As the product is made of resin and comes in a variety of colors, we used this to benefit the design. This execution allows the packaging to take on a variety of colors based on the product inside, and at the same time, urge the viewer to open the container. It is perfect as a gift box as well.

4. Genki-Aroi

Client Name:	Alpine Foods/Thailand
Brand Name:	Genki & Aroi
Type of Product:	Seasoning mixes
Material:	Art Card paper
Theme or Concept:	Afer succeeded with Genki - Japanese seasoning mixes, the client launched a new brand, Aroi - Thai seasoning mixes. The packaging followed Genki design format with Thai essence and motifs.

5. Genki

Client Name:	Alpine Foods/Thailand
Brand Name:	Genki
Type of Product:	Seasoning mixes
Material:	Art Card paper
Theme or Concept:	A packaging design for Genki, a line of healthy and delicious Japanese seasoning mixes. The name "Genki", which means healthy in Japanese, provides the main selling point for the product. The logo and the packaging have a fun and lively look to gain the attention of both adults and children.

6

7

6. In2it Pure & Natural

Client Name:	In2it Pure & Natural/Thailand
Brand Name:	In2it Pure & Natural
Type of Product:	Cosmetics
Material:	Art Card paper
Theme or Concept:	Pure & Natural is a series of cosmetics from in2it, which contain ingredients including fruit and vegetable extracts. The idea is to actually see slices of these fruits and vegetables in the products through transparent packaging.

7. Alternative Hits CD Album

Client Name:	EMI/Thailand
Brand Name:	EMI
Type of Product:	CD Album
Material:	Plastic
Theme or Concept:	A packaging design for a compilation of alternative music hits created for EMI.

8. Ashford Eye Drops

Client Name:	Ashford Laboratories/Macau
Brand Name:	Ashford Eye Drops
Type of Product:	Eye Drops
Material:	Art Card paper
Theme or Concept:	The Ashford eye drops packaging was redesigned to improve the brand image. The assignment was to create a package that had a professional medical look, yet still appealed to ordinary optical customers and could be displayed on shelves either vertically or horizontally.

8

243

DOLHEM DESIGN
SWEDEN

Dolhem Design, create, refine and clarify our clients' brands, products and services through a mixture of quality, creativity, charisma and economical thinking (in the order preferred by the client). Even is we use different models depending on client need and objectives, design management is always the common denominator.

We have built up a broad client base that spans a large range of different businesses in both the public and private sectors since 1998. As a result of belonging to INAREA, an international design network, we have access to expertise in different markets and an inexhaustible source of information that can be used on behalf of our clients. Dolhem Design's international character can also be seen in the fact that five languages are fluently spoken at the company (Swedish, English, French, and Czech).

Feel free to order our case stories or to book a meeting with us so that we can explain in detail how we work and look upon branding as well as how design management can be seen as an investment rather than only as an expense.

Please visit: www.dolhemdesign.se

Dolhem Design
Storgatan 22 A
SE-11455 Stockholm
Sweden

Tel: +46 (0)8 661 50 47
Fax: +46 (0)8 661 50 48

info@dolhemdesign.se
www.dolhemdesign.se

Akademikliniken
Dolhem Design produced a series of medical skin care products in its work for Akademikliniken. Softly formed bottles and jars that allow the colours of the products to show through and luxurious white cartons are included in the range. The natural colours of the products have been a source of inspiration for Dolhem Design's work with Akademikliniken.

The skin care series can be seen as a marketing instrument for Akademikliniken in that it spreads awareness of the brand as well as related positive values to a broader target group.

Akademikliniken is one of the world's largest private hospitals for aesthetic and reconstructive plastic surgery. Dolhem Design has worked with Akademikliniken during the clinic's vigorous growth period and formed the basis for its brand profile and design concept.

EICHE, OEHJNE DESIGN
GERMANY

Ilka Eiche and Peter Oehjne, founder and owner of the agency

Most markets are so competitive that the marketing departments of companies must act with utmost flexibility and a highly professional attitude. With our core competence – Corporate Design and Packaging Design – we think that we can offer instruments to meet the demands of the market. We believe Corporate Design and Packaging Design are an extension of brand development. For many people, the word "corporate" describes a static system – one that is inflexible and difficult to change. We believe, however, that corporations are dynamic systems. This defines our approach to the market and design. That's how corporations stay flexible and fast-moving.

EOD offers a full range of services to execute Corporate Design and Packaging Design projects, and these services come with personal involvement and innovative thinking. This includes consulting, strategy and conceptualization – from design to the end product. In the process of designing and implementing projects, we are always striving for distinctive, progressive and high-quality solutions that distinguish our clients from the masses. In short – our goal is to combine systematic thinking with first-class design. We have been rewarded for our approach with long-term business relationships and international prizes. Some of our leading ideas are:

- Steer toward a unique, brand-typical appearance on the market
- Increase visibility via consistent and progressive brand design
- Raise customer loyalty through precise definition of the target audience
- Improve turnover through high-quality design
- Raise the value of the company through increased brand value

Since we were founded in 1999, we have worked for numerous companies, institutions and associations. Both mid-sized service providers and large corporations.

Fritz Allendorf Winery, The Gold Line

EICHE, OEHJNE DESIGN / GERMANY

FRITZ ALLENDORF WINERY
THE KLASSIK LINE
WINE
GLAS AND PAPER
KLASSIK LOOK

FRITZ ALLENDORF WINERY
SPECALITIES
SPARKLING WINE
GLAS AND PAPER
EMOTIONALITY

EICHE, OEHJNE DESIGN / GERMANY

FRITZ ALLENDORF WINERY
THE EMOTION LINE
WINE
GLAS AND PAPER
HAVE FUN

FRITZ ALLENDORF WINERY
THE EMOTION LINE
SPARKLING WINE
GLAS AND PAPER
HAVE FUN

EICHE, OEHJNE DESIGN / GERMANY

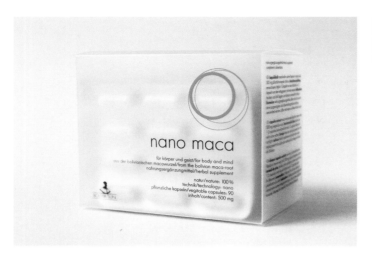

ECCO TERRA
NANO MACA
HERBAL SUPPLEMENT
POLYPROPYLEN & BLISTER
REDUCEMENT

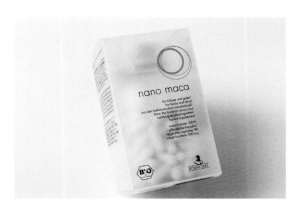

ECCO TERRA
NANO MACA
HERBAL SUPPLEMENT
POLYPROPYLEN & BOTTLE
REDUCEMENT

BRAND & COMPANY

KOREA

Founded in 1986, providing Total Brand Consulting Services including Brand Strategy, Brand Research, Brand Naming, Brand Design and Brand Communication

404 Sangjeok-Dong, Seongnam-Si,
Sujeong-Gu, Gyeonggi-Do, South Korea
Tel: 82.2.578.2889 Fax: 82.2.577.3666
www.brandconsulting.co.kr

Contact :
katy choi
e-mail : katy@brandconsulting.co.kr

1. LOTTE, South Korea
2. Drip Coffee
3. Coffee
4. Tea Bag / Cardboard
5. Emphasis on rich aroma, creating the images for each flavor

1. LOTTE, South Korea
2. Arbes
3. Olive Oil
4. Label Graphic Design / Sales Kit
5. Communicating the heritage and high Quality of Premium Olive Oil

1. LOTTE, South Korea
2. BBQ Olive Oil
3. Olive Oil
4. Label Graphic Design / Sales Kit
5. Communicating the high Quality of Olive Oil

1. LOTTE, South Korea
2. ILAC
3. Soft Drink
4. Pet Bottle with Printed Label
5. New family brand of low-carbonated soft drink for teens and young adult

1. Muhak, South Korea
2. Good Day
3. Reduced Alcohol Beverage
4. Label Graphic Design / Sales Kit
5. The calligraphy and the illustration of nature reflecting traditional Korean design

WILLIAM FOX MUNROE, INC
U.S.A.

William Fox Munroe has been developing graphic design for packaging for over 35 years. We opened in 1972 as a one-man, one-client design shop. Things grew from there, and in 1998 things began to change when employees Dan Forster, Tom Newmaster, and Steve Smith purchased the business and began to focus exclusively on package design and creating a more compelling experience for the consumer at point of sale.

In the years since they purchased the agency, the partners have put programs into place that have helped our agency take off. In a time when others in the communications community experienced round after round of layoffs, William Fox Munroe more than doubled staffing and more than tripled work volume.

In 2005, WFM was voted #16 in Fast Company Magazine's Fast 50. This list highlights the 50 most innovative companies in the world.

What's our secret to all this success? Differentiation.

We're focused. That means we only do packaging and packaging support. No paid media. Just package design 24/7.

We're fast. When you live in the packaging world, you understand that it doesn't matter if it's good if it's just too late. Deadlines are critical.

We're user-friendly. No attitudes or egos here. We make sure you never feel like we're doing you a favor by completing your project on time and within budget.

And finally, it's much easier to be successful when you love what you do. We care about what we do and it shows. And we plan to be loving our work for many years to come.

The Hershey Company / USA
Hershey's Sticks
Confectionery
William Fox Munroe, Inc.

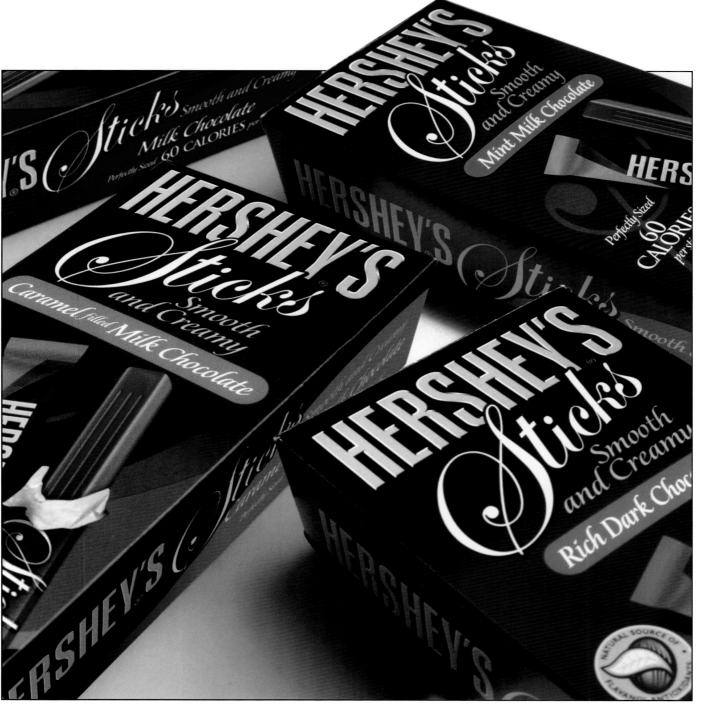

The Hershey Company

1
The Hershey Company / USA
Hershey's Syrups
Food and Beverage
William Fox Munroe, Inc.

2
The Hershey Company / USA
IceBreakers Unleashed Limited Edition
Gum and Mint
William Fox Munroe, Inc.

3
The Hershey Company / USA
IceBreakers Limited Edition
Gum and Mint
William Fox Munroe, Inc.

4
The Hershey Company / USA
Valentine Packaging
Confectionery
William Fox Munroe, Inc.

5
The Hershey Company / USA
Character Stocking Stuffers / Gift Toppers
Confectionery
William Fox Munroe, Inc.

6
The Hershey Company / USA
Reese's Nutrageous
Confectionery
William Fox Munroe, Inc.

1
R.M. Palmer Company / USA
Choceur Easter Packaging
Confectionery
William Fox Munroe, Inc.

2
Murry's / USA
Pizza Design
Frozen Foods
William Fox Munroe, Inc.

3
The Silver Palate / USA
The Silver Palate Salad Splashes
Gourmet Foods
William Fox Munroe, Inc.

4
Bed, Bath and Beyond / USA
Stainless Steel Product Line
Kitchen Accessories
William Fox Munroe, Inc.

5
W. L. Gore & Associates, Inc. / USA
Glide Dental Hygienist Kit
Promotional
William Fox Munroe, Inc.

6
Lowe's Companies, Inc. / USA
Real Organized Product Line
Home Organization
William Fox Munroe, Inc.

COBA & ASSOCIATES

SELBIA

Coba&associates
Creative Business Excellence

Coba&associates is design and branding agency. Our core competencies are corporate branding, packaging design, digital branding and all services connected to it.

We believe that design, as widely defined discipline, can upgrade communication and be used as strategic tool for succesfull business and culture.

Branding is everything to which a company should aspire.
It helps growth, control of property, innovation. It's a customer's window to a new and exciting product. If it excites it lives. Smart branding makes product live longer.

Design works as predilection of future events. It communicates on different levels and delivers stability in how you talk, look and behave. It gives power to a larger group of people connected by a common interest. It gives opportunity and time to think and act properly. It's entirely dedicated to what its going to be, not what has already happened.

Coba&associates
Carice Milice 3,
11000 Belgrade, Serbia

tel/fax: +381.11.30.34.900
office@cobaassociates.com

www.cobaassociates.com

1. Foodex
   Nature's Plan
   Fruit Preserves
   Small batch production reminiscent of organic natural product

2. CARLSBERG
   Lav Beer
   Premium Lager
   On-trade brand

3. CARLSBERG
   Lav Beer
   Lager
   Off-trade brand

4. TIKVES
Reserve Wine
Red and white wines range
Premium labels for quality Macedonian wines

5. CARLSBERG
Cortez
Flavored beer
Cutting edge taste for young and hip deserved
sharp brand image

7. RAJKOVIC SPRINGS
   Raj Water
   Mineral water
   Peice of heaven in your hand

8. VALLETTA DAIRY
   Mozzarella
   Cheese product
   Trustworthy, clean

9. FILIP MCM
   River Fish
   Canned fish
   Natural, straightforward
   back-to-basics touch

8.

BMU CREATIVE DESIGN AGENCY

TURKEY

Salih Küçükağa / Chief Creative Officer

BMU is a creative design agency with years of experience; creating brand identity, packaging, print media, and interactive designs for a wide range of cultural and corporate clients. Whether on screen, print, or in strategic brand development, successful solutions require a balance of innovation and pragmatism based on knowledge, experience, and the drive to do more. BMU, strikes this balance in its foundation from which we operate.

Capabilities

BMU is focused to achieve implement and achieve the best possible marketing strategy for you. We are confident in all our ideas and will not be satisfied unless you are. From the start until the end of the project phase, we are totally committed and fulfill our tasks to the utmost requirement. Every client is unique and we strive to create the solution that is unique to you. Our aim with each project consists of the right branding, placement, aimed at the right audience, sufficient planning, successful management, precise mediums used and most importantly, ideas that grow.

What we strive for?

At BMU we strive to push the boundaries of design and print media. We understand that in order for visual information to be communicated effectively it must possess sound, smart and innovative design principles.

"Defining the idea, identifiying the brand and finally creating the identity. Are the key principals we go by in order to push the boundries of visual communication."

1. Nazo Hot Chocolate

    Client: Nazli
    Type of Product: Instant Hot Chocolate
    Services: Package Redesign, Product Strategy & Positioning, Form & Structure
    Objectives: Refreshing the look of Nazo Hot Chocolate 200ml.

2. Flipper Fruit Flavoured Powder Drink

    Client: Nazli
    Type of Product: Fruit Flavoured Powder Drink
    Services: New Product Packaging, Category Design Audit, Form & Structure
    Objectives: Developing brand identity and bilingual packaging for a new brand of European instant beverages company.

3. Nazo Instant Coffee (3in1)

    Client: Nazli
    Brand Name: Nazo
    Type of Product: Instant Coffee
    Services: Package Redesign, Product Strategy & Positioning, Form & Structure
    Objectives: Redesigning the package of Nazo Instant Coffee (3in1) with clean, understated graphics.

4. Gelin Liquid Soap

    Client: Omen Cosmetic
    Type of Product: Liquid Soap
    Services: Brand developement, Brand & Product Strategy, Form & Structure, Product Innovation
    Objectives: A liquid soap that is commonly found within our bathrooms with a floral and fruity essence is under question of "how it can distinguished from its competitors, and transformed to project a decor object within our homes?"

1. Nazo Hot Chocolate

2. Flipper Fruit Flavoured Powder Drink

2. Flipper Fruit Flavoured Powder Drink

4. Gelin Liquid Soap

3. Nazo Instant Coffee (3in1)

4. Gelin Liquid Soap

TD2, IDENTITY AND DESIGN CONSULTANTSANTS

MEXICO

Rafael Treviño / President
Rafael Rodrigo Córdova / Creative Directo r.

Established in 1886, as Treviño Diseño, started opertartions with clients like Procter & Gamble, Sabritas, AC Nielsen today is one of the most important independent Design companies in México Cit y.
The firm is leaded by Rafael Treviño / President and Rafael Rodrigo Córdova / Creative Director.
It changed its name to TD2 in 1993 and since thenm TD2 has grown eve ry year to develop Identity projects for consumer goods -locally and regionally- for clients such as Bachoco, Bimbo, Coca-Cola, LG electr onics, Motorola, Nestlé, Novartis.

Building up a brand identity in peoplés minds is harder eve ry day, but the benefits for the companies that achieve it are huge.
These days it is a must to be a dynamic and efficient compan y, which can respond quickly to the market changes, and to the constant evolution of the consumer habits in today's world.
The really important thing is to make people have a perception of yours brands as their best choice, the ones which come closest to their desires and emotions, those which best fit their needs and lifestyles.
TD2 plans, designs and executes visual communication systems to invigorate performance of parameters which generate brand value perception.
TD2's specialized communicators and designers define a brand 's strategic course together with our clients, allowing us to come up with results that are innovative, relevant for the targeted consumer, brilliant in execution and with a high level of quality through all the visual elements in contact with clients and suppliers.
TD2 helps its client's business performance by ensuring their brands are perceived as alive, contemporary, relevant and close to their consumers in emotional terms.

1. Tocamba, Guadalajara México
2. Tequila Insignia
3. Tequila
4. Glass, Paper
5. Traditional mexican tequila

www.td2.com.mx
contacto@td2.com.mx

1. Nestlé Nutrition, México
2. Nestum Bienestar
3. Infant cereal
4. Metallic paper
5. Packaging for a premium infant cereal

1. Nestlé México
2. Fiorelo cream cheese
3. Cheese
4. Card stock
5. Packaging for a cream cheese

1. Casa San Matías, Guadalajara México
2. Tequila Pueblo Viejo
3. Tequila
4. Glass, paper
5. Packagin for a premium Tequila

1. Bachoco, Celaya México
2. Consomé de Pollo Bachoco
3. Chicken broth
4. Pape r, card stock
5. Packaging for a local chicken producer

1. Nestlé México
2. Carlos V chocolate
3. Chocolate
4. Paper
5. Packaging for a local chocolate brand

1. Nestlé México
2. Larín chocolate
3. Chocolate
4. Paper Foil
5. Packaging for a local chocolate brand

1. Casa San Matías, Guadalajara México
2. Tequila San Matías
3. Tequila
4. Glass, paper
5. Packaging for a TraditionalTequila

1. Diageo, México-Scotland
2. Buchanan 's De Luxe
3. Whiskey
4. Tin Can
5. Conmemorative packaging

1. Cereal Partners Worldwide, México
2. Golg cereal
3. Cereal
4. Card stock
5. Conmemorative packaging

281

1. Nestlé, México
2. Helados Nestlé
3. Ice cream
4. Plastic
5. Packaging for ice cream (6 inks)

1. Nestlé, México
2. Galáctea 7
3. Creamsicles
4. Foli
5. Packaging for a creamsicles for kids

1. Nestlé, México
2. Nestlé Fruit yogurt
3. Yogurt
4. Paper
5. Packaging for Nestlé Yogurt

1. Naturitz, México
2. Bee suite
3. Powdered honey
4. Card Stock
5. Packaging for powdered honey

1. Nestlé Nutrition México
2. NAN Pro
3. Infant formula
4. Foil
5. Packaging for an infant formula.

1. Colgate, México
2. Freska-ra
3. Tooth paste
4. Card stock
5. Packaging for a local toothpaste

noAH is an organic package design directory for your international creative business needs.

Within this book there are support systems for art coordination projects to enhance your domestic or global commerce.

ICO began supporting international business communication for clients and creators with noAH-7. We wish to continue our support in noAH-8, and in the future.
The index for the following pages may be found in the ICO site.

http://www.1worldart.com/

From the ICO site you may jump directly to each participating designer's homepage.

ICO international art coordinators can support you worldwide.
In advance of your project ICO can assist in feeling out the designer's work possibilities, and approximate their budget.

The main function is :

1 ICO can nominate recommended artists for your project, upon your request.
2 ICO can research the designer's time schedule.
3 ICO can assist you in researching the fit of your project's budget with the designer in advance.
   In addition, ICO can assist to negotiate the project's budget with both parties.
4 ICO can support each party's communication problems up to you'll start the production stage.

ICOでは、クライアンツとクリエイターズの国際的アート・ビジネスのサポートを開始しました。
本書「世界のパッケージ年鑑」"noAH"は、クリエイティブなお仕事に従事されている方々が、日々利活用できる生きた書籍です。

ICOのサイトには、本書と同じIndexページがアップロードされています。そのサイトのアイコンをクリックすれば、各作家のホームページを直接訪問する事ができます。

http://www.1worldart.com/

また、言葉の問題や、煩わしい価格交渉ごと、作家のタイムスケジュール調整など、ご希望があれば、ICOの国際アート・コーディネーターがいつでもサポートします。
どうぞお気軽にご利用下さい。

このプロジェクトの主な機能は：

1 各作家のタイムスケジュール・リサーチ

2 事前の作家制作予算リサーチ、及び、予算価格調整

3 プロジェクトが具体的な制作に掛かるまで、需給相互間のコミュニケーション・サポート

4 また、もし言葉の障害でお困りの方には、実作業がスタートされるまでの期間、ICOの国際部が双方のコミュニケーション・サポートを致します。

詳しくは、お気軽に下記のICOへお問い合わせ下さい

For more information go to :
ICO Headquarters : Norio Mochizuki <icohq@info.email.ne.jp>
ICO Japan : Satoru Shiraishi <c-wave@gaea.ocn.ne.jp>
ICO America : Robert Morris <icoamerica@cox.net>
ICO America / Kansas : Jo Ssickbert <Josanartist@aol.com>

INDEX

This index page synchronizes with each designer's home page via ICO's website. Please click the following ICO site/ so that you can visit their fantastic studio and outstanding works worldwide.
If you have any inquiry or language problems, please contact ICO headquarters freely. All ICO groups will suport your art coordination according to your specifications.

STEP-1　http://www.1worldart.com/
STEP-2　http://www.1worldart.com/aa/jb/category_menu.html
STEP-3　http://www.1worldart.com/aa/jb/first_page_package_d/package_cover.html
STEP-4　http://www.1worldart.com/aa/jb/first_page_package_d/search_by_designer/search_by_designer.html

	Barrie Tucker P.O. Box 390 Nairne SA 5252 Australia	P42-49
	BETC DESIGN 19.23 Passage du Desir 75010 Paris, France	P62-69
	BENEDICT CORPORATION 8F 2-11-8 Meguro Meguro-ku Tokyo 153-0063 Japan	P96-99
	B & G (Black & Gold) 82, avenue Marceau 75008 Paris, France	P160-165
	Brandaid Luteranskaya str. 33, office 1 Ukraine	P238-239
	Brandnewdesign Leeuwenveldsweg18, 1382LX Weesp, Postbus 289, 1380 AG Weesp, The Netherlands	P36-41
	Brand & Company 404 Sangjeok-Dong, Sujeong-Gu, Seongnam-Si, Gyeonggi-Do 461-320 Korea	P256-259
	BMU CREATIVE DESIGN AGENCY Kisikli Mh. Abdipasa Sk. 29/12 Uskudar 34672 Istanbul, Turkey	P272-277
	Cato Partners 10 Gipps St. Collingwood, Victoria 3066 Australia	P90-95
	Coba & Associates Carice Milice 3, 11000 Beograd, Serbia	P266-271
	Crave, Inc. 3100 NW Boca Raton Blvd. Suite 109 Boca Raton, FL 33431 USA	P 192-197
	DESIGNAFAIRS 3100 NW Boca Raton Blvd. Suite 109 Boca Raton, FL 33431 USA	P144-145
	DEPOT WPF 109004, Moscow, Pestovsky pereulok, 16,bld.2, Business CenterAKMA, 4th fl., Russia	P146-153
	Deutch Design Works 10 Arkansas St. San Francisco, CA 94107 USA	P82-89

このインデックスページは、ICO のウェブサイトで各デザイナーのホームページにリンクしています。以下の ICO のホームページから、それぞれの作家のイメージ・アイコンをクリックしてみて下さい。世界中のファンタスティックな彼らのスタジオや、素晴らしい作品の数々を訪ねることができます。

もし質問やお問い合せ、言葉の問題などがございましたら、いつでも ICO 本部にご連絡下さい。あなたの要請に従って、各国の ICO グループがプロジェクトに関するアート・コーディネーションのお手伝いを致します。

STEP-1　http://www.1worldart.com/
STEP-2　http://www.1worldart.com/aa/jb/category_menu.html
STEP-3　http://www.1worldart.com/aa/jb/first_page_package_d/package_cover.html
STEP-4　http://www.1worldart.com/aa/jb/first_page_package_d/search_by_designer/search_by_designer.html

DIL BRANDS
Andres Bello 2777 Of. 2403 Las Condes, Santiago, Chile
P100-107

Dolhem Design
Storgatan 22 A, 114 55 Stockholm, Sweden
P244-247

Eiche, Oehjne
DeaignKaiser-Friedrich-Promenade 21, 61348 Bad Homburg bei Frankfurt, Germany
P248-255

FiF DESIGN
21/89 Soi Soonvijai, Rama 9 Rd., Bangkapi, Huaykwang, Bangkok 10310 Thailand
P220-225

Graham Hanson Design
60 Madison Ave. Floor 11, New York, NY 10010 USA
P172-173

IFF COMPANY INC.
6-8-11 Minami-aoyama Minato-ku Tokyo Japan
P124-129

INGALLS+DASSOCIATES
10 Arkansas St. Studio E San Francisco, CA 945107 USA
P22-29

im2design (Monika kaczmarek)
ul. Redowa 11, 93-134 Lodz, Poland
P210-213

Joao Machado design Lda
Rua Padre Xavier Coutinho, 125 Portugal
P14-21

Kazumi Kagawa Design Office
5-11-501-1503 Koyocho-naka, Higashinada-ku Kobe 658-0032 Japan
P2138-143

KEIZ OBJ WORKS INC.
6-20-3 Chuoh Ohta-ku Tokyo 143-0024 Japan
P186-191

khdesign
Lilistrabe 83D/09, 63067 Offenbach, Germany
P30-35

Midnite Oil Design
5/38 Plus 38, Sukhumvit 38 Rd., Phrakhanong, Klongtoey, Bangkok 10110, Thailand
P240-243

MILK
8 Bakopoulou str. 15451 N. Psihiko, Athens, Greece
P166-169

INDEX

Minato Ishikawa Associstes inc. — P154-159
4-12-7-503, Nakameguro, Meguro-ku, Tokyo 153-0061 Japan

Mountain Design — P76-81
Scheveningseweg 42, 2517 KV The Hague, The Netherlands

ONC DESIGN STUDIO — P214-219
3-2-17-214 Nakazaki Nishi Kita-ku, Osaka 530-0015 Japan

PATRICIA CATALDI — P184-185
Praca Vilaboim 78 ap 6 Andar Higienopolis 01241-010 Sao Paulo / SP Brazil

pearlfisher — P236-237
50 Brook Green London W6 7BJ United Kingdom

Stromme Throndsen DESIGN AS — P70-75
Holtegata 22, 0355 Oslo, Norway

STORM CORPORATE DESIGN LTD — P226-231
1-154 Hendon Ave., Mount Albert, Auckland 1025 New Zealand

STUDIO GT&P — P118-123
Via Ariosto, 5, 06034 Foligno, Italy

STUDIO360 — P174-177
Kotnikova 34, SI-1000 Ljubljana, Slovenia

Tangram Strategic Design — P206-209
viale Michelangelo Buonarroti 10/C, 28100 Novara, Itlay

TD2, IDENTITY AND DESIGN CONSULTANTS — P278-283
Ibsen 43 8 Piso, Col. Polanco, D.F. 11560 Mexico

the LemonYellow Design — P58-61
9F-3, No. 512, sec.4, Chung-Hsiao E. Rd. Taipei, Taiwan

tridimage — P130-137
Av. Congreso 4607, C1431AAB Buenos Aires, Argentina

Triple888 — P178-138
83 Wigram St. Parramatta NSW 2150 Australia

ICO HQ. / Norio Mochizuki
< ico-nori@info.email.ne.jp >
ICO AMERICA California / Robert Morris
< ICOAmerica@cox.net >
ICO AMERICA Kansas / Jo Sickbert
< JOSICKBERT@aol.com >
ICO JAPAN / Satoru Shiraishi
< c-wave@gaea.ocn.ne.jp >

TUCKER CREATIVE 57c Kensigton Road, Norwood SA 5067 Australia		P50-57
UNIQA.C.E 105062, Pokrovka, 30, Moscow, Russia		P198-205
UP CREATIVE No. 54, Ching-cherng St., Taipei, Taiwan		P112-117
Vie design Ciateul Tengah 7, Bandung 40252 Jawa Barat, Indonesia		P232-235
VILLEGERSUMMERSDESIGN 29 Blenheim Gardens London NW2 4NR United Kingdom		P170-171
WILLIAM FOX MUNROE, INC 3 E. Lancaster Ave., Shillington, PA 19607-0065 USA		P260-265
Yashi Okita DESIGN 1111 Howard Ave., Burlingame, CA 94010 USA		P108-111
Prologue	Jo Sickbert	P 1
Essay	Yashi Okita	P 2-3
Essay	Anna Lukanina	P 4-5
Essay	Jeffrey Su	P 6-7
Essay	Barrie Tucker	P 8-9
Essay	Mike Branson	P 10-11
Essay	Faith Dennis Morris Ed.D	P 12-13

PENTAWARDS

Packaging designers now have their Worldwide Packaging design Competition!

Pentawards is the first and only worldwide online competition exclusively devoted to packaging design in all its forms. It is open to everybody in all countries who are associated with the creation and marketing of packaging. The winners will receive bronze, silver, gold, platinum or diamond Pentawards according to the creative quality of their work.

Creations from the world over will be judged by a jury of 12 pro's, itself international, who will select the winners in accordance with the creative quality of the work submitted.
Five major categories and 40 sub-categories will allow a large range of creations to participate.

Apart from prize-giving, Pentawards' mission is the promotion of packaging design with companies, the press, the economic and political authorities and the public in general, throughout the world.

By participating in Pentawards, packaging designers will have the opportunity of comparing their creations with others from all over the world, and they will have the possibility of winning a prestigious award which will allow them to demonstrate their creativity and expertise.

■ The trophies:

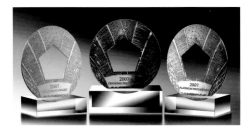

Gold, Platinum and Diamond will receive the prestigious Pentawards
Trophies. They are the work of the American artist Christian Heckscher
(http://www.liftdesignetching.com).
The Diamond Pentaward ("Best of the Show") will be adorned by an inlaid,
authentic 1 carat diamond with a value of over 2500Euros, certified by the diamond bourse of Antwerp in Belgium.
The founders:

Jean J. and Brigitte Evrard, partners in life and in business for more than 35 years; designers, they founded their packaging design agency in 1976. Their agency became Carre Noir Brussels from 1994 to 1998.
They then joined the international group Desgrippes Gobe Enthusiasts about supermarkets the world over, about packaging and convenience goods, they have the credit of having organised wonderful FMCG exhibitions: Shopping in Tokyo (1995 - 2003 - 2006), Shopping in Cartoon (1996), Shopping in China (2000), Shopping in Europe (2006), which attracted thousands of visitors.

Together they have won more than 50 awards. They have now decided to devote themselves to the worldwide promotion of a communication tool to which they have given their lives: packaging.

The creation of Pentawards, a world packaging-design competition, shows that they consider packaging as one of the foremost communication tools in global business and they intend to make this known to the profession and the press.

www.pentawards.org question@pentawards.org

Should you need more : then you may add the list of the categories:

■ Major categories and sub-categories:

■ BEVERAGES
* Water
* Soft drinks, juices
* Tea and coffee (RTD - ready-to-drink)
* Milk, chocolate (RTD)
* Sport, vitamin & energy drinks
* Beer, cider and low-alcohol or non-alcoholic drinks
* Wine, spirits
* Beverage trends (diet, weight control, health beverages, organic, fair trade etc.)

■ FOOD
* Cereals (cereals, bread, pasta, soya, rice, noodles, dried pulses etc.)
* Coffee & tea (dry)
* Dairy products (cheese, yoghurt, eggs, butter, margarine etc.) excluding beverages
* Spices, oils and sauces (oil, vinegar, mayonnaise, mustard, spices, sauces etc.)
* Fish, meat, poultry (fresh, deep-frozen, dried or canned)
* Fruit and vegetables (fresh, deep-frozen, dried or canned)
* Soups, ready-to-eat dishes (fresh, deep-frozen, dry or canned)
* Confectionery and sweet snacks
* Savoury snacks
* Pastry, biscuits, ice-cream, desserts
* Food trends (diet, weight control, health food, organic, fair trade etc.)

■ BODY
* Clothing (clothing, shoes, underwear, hosiery, haberdashery,...)
* Health care (OTC, pharmaceuticals, vitamins, optical, toilet paper, tissues, feminine hygiene, diapers etc.)
* Body care (soaps, bath & shower products, deodorants, dental products, hair care, shaving, sun protection, hand and foot care etc.)
* Beauty (perfumes, cosmetics, make-up, body decoration etc.)

■ OTHER MARKETS
* Household maintenance (detergents, cleaning products and utensils, batteries, light bulbs etc.)
* Home improvement (paint, tools, gardening products, D.I.Y. etc.)
* Electronic (computers, printers, consumables, software etc.)
* Non-electronic (paper, writing materials, stationery etc.)
* Automobile products (additives, oils, car care, spare parts etc.)
* B2B products (all types of business)
* Brand Identity Programs
* Distributors/Retailers own brands (range of minimum 10 SKU's)
* Pet Products (food and accessories)
* Entertainment (sport, music, photography, phones, games, toys etc.)

■ LUXURY
* Perfumes
* Make-up
* Spirits
* Fine wines, champagne
* Cigars & tobacco
* Jewellery, watches, fashion garments
* Gourmet food

Epilogue

We have been surprised recently that so many Japanese enterprises are using many overseas designers for their assignment projects of Japanese products.

Riding on the wave of globalization, the number of commodities in Japan are making inroads into foreign markets, and it is increasing every year.

Conversely, the products from foreign countries spreads outside its borders, and are beginning to infiltrate Japan. To request a new visual sense, even a domestic-oriented commodity begins to appoint a foreign package designer resolutely along with it to transform itself.

With the advance of this fact, even in Japanese products for the domestic Market, they are using overseas designers positively, in order to get a new visual sense. This globalization has infiltrated the general consumption market. In this age, if someone said that the image of Japan is characterized by the samurai and geisha girl it would be a ridiculous statement. We cannot succeed in the marketplace if we haven't the understanding and recognition of the correct market in each country.

In this phenomenon, perhaps, both seem to be the same in the tendencies of every country in the world. When future books of this genre will be compiled in the future, I will bear in mind that ICO must analyze the consumer trends of the world, and search for excellent work suitable for these needs.

We wish to express our gratitude for the pleasant cooperation and support of the following people who have helped during the publication of this book.

Anna Lukanina,
Barrie Tucker,
Faith Morris,
Jeffrey Su,(LemonYellow Design)
Jennifer Morris,
Leon Bosboom,
Margarethe Hubauer,
Mike Branson,
Robert Draper,
Satoshi Yonekura,
Slobodan Jovanovic,
Yashi Okita,
Zhu Gang,

ICO noAH-8 Compilation Project

Norio Mochizuki, Robert Morris, Jo Sickbert,
Yumiko Mochizuki, Satoru Shiraishi,

■ **Title**
World Package Design noAH - 8
■ **Released**
January. 2008
■ **Art Supervision**
Yumiko Mochizuki, Jo Sickbert, Robert Morris,
Colette Coltte, Satoru Shiraishi
■ **Cover Design**
Norio Mochizuki, Kazuhito Mochizuki
■ **Cover images**
Brandnewdesign, Coba & Associates
■ **Editorial Design Layout**
Yumiko Mochizuki, Kazuhito Mochizuki
■ **Translation**
Robert Morris, Jo Sickbert
■ **Technical support**
Kazuhito Mochizuki, Satoshi Yonekura,
Robert Draper
■ **Publishing House**
ICO CO., LTD. Publishing House,
ICO (International Creators' Organization)
Post Code 253-0051 13-14 Wakamatsu-cho,
Chigasaki, Kanagawa, JAPAN
t : 0467-87-2110   f : 0467-86-8944
url : http://www.1worldart.com/
■ **Publisher**
Norio Mochizuki
■ **ICO site**
http://www.1worldart.com/
■ **noAH INDEX site**
http://www.1worldart.com/aa/jb/first_page_
package_d/package_cover.html

ICO Public Art Production project :

ICO HQ. / Norio Mochizuki
< ico-nori@info.email.ne.jp >
ICO AMERICA California / Robert Morris
< ICOAmerica@cox.net >
ICO AMERICA Kansas / Jo Sickbert
< JOSICKBERT@aol.com >
ICO JAPAN / Satoru Shiraishi
< c-wave@gaea.ocn.ne.jp >
ICO FRANCE / Colette Cotte
< colette.desseigne@tele2.fr >
ICO ITALY / Elisa Vladilo
< elisabrzo2003@yahoo.co.uk >

ICO Distribution Activities :

ICO HQ. (*INTERNATIONAL CREATORS' ORGANIZATION*)

13-14 Wakamatsu-cho, Chigasaki,
Kanagawa, JAPAN Postal code : 253-0051
t   : 81(Japan) (0)467-87-2110
f   : 81(Japan) (0)467-86-8944
e   : <ico-nori@info.email.ne.jp>

ICO Distribution & Book Promotion Center

Contact to :
Robert Morris in ICO AMERICA.
< ICOAmerica@cox.net >

タイトル / World Package Design noAH - 8
発行日 / 2008年1月
表紙デザイン / 望月 紀男 / 望月 かずひと
編集制作 / 望月 由美子、ジョー・シックバート /
ロバート・モーリス、望月 かずひと、白石 智
本文レイアウト デザイン構成 / 望月 由美子、望月 かずひと
翻訳 / ロバート・モーリス、ジョー・シックバート
発行所 / ㈱ ICO
〒253-0051 神奈川県茅ヶ崎市若松町13-14
t : 0467-87-2110   f : 0467-86-8944
url : http://www.1worldart.com/

ICOサイト :
http://www.1worldart.com/
noAH INDEX サイト :
http://www.1worldart.com/aa/jb/first_page_package_d/
package_cover.html

発行人 / 望月 紀男 / ICO HQ.

Distributor : AZUR corporation
1-44-8 Jinbou-cho, Kanda, Chiyodaku, Tokyo
t : 03-3292-7601   f : 03-3292-7602

定価 : 9,500 yen

ISBN 978-4-931154-28-5C3072-9500E

DATE DUE	RETURNED
NOV 0 1 2011	NOV 1 2 2011
MAR 1 5 2012	MAR 1 0 2012
JUL 1 7 2013	JUL 3 0 2013

RECEIVED

JUL 2 1 2011